Renewing Destruction

TRANSFORMING CAPITALISM

Series Editors

Ian Bruff, University of Manchester; Julie Cupples, University of Edinburgh; Gemma Edwards, University of Manchester; Laura Horn, University of Roskilde; Simon Springer, University of Newcastle; Jacqui True, Monash University

This book series provides an open platform for the publication of path-breaking and interdisciplinary scholarship which seeks to understand and critique capitalism along four key lines: crisis, development, inequality and resistance. At its core lies the assumption that the world is in various states of transformation, and that these transformations may build upon earlier paths of change and conflict while also potentially producing new forms of crisis, development, inequality and resistance. Through this approach the series alerts us to how capitalism is always evolving and hints at how we could also transform capitalism itself through our own actions. It is rooted in the vibrant, broad and pluralistic debates spanning a range of approaches which are being practised in a number of fields and disciplines. As such, it will appeal to sociology, geography, cultural studies, international studies, development, social theory, politics, labour and welfare studies, economics, anthropology, law and more.

Titles in the Series

The Radicalization of Pedagogy: Anarchism, Geography, and the Spirit of Revolt. Edited by Simon Springer, Marcelo de Souza and Richard J. White

Theories of Resistance: Anarchism, Geography, and the Spirit of Revolt. Edited by Marcelo Lopes de Souza, Richard J. White and Simon Springer

The Practice of Freedom: Anarchism, Geography, and the Spirit of Revolt. Edited by Richard J. White, Simon Springer and Marcelo Lopes de Souza

States of Discipline: Authoritarian Neoliberalism and the Contested Reproduction of Capitalist Order. Edited by Cemal Burak Tansel

The Limits to Capitalist Nature: Theorizing and Overcoming the Imperial Mode of Living. Ulrich Brand and Markus Wissen

Workers Movements and Strikes in the 21st Century. Edited by Jörg Nowak, Madhumita Dutta and Peter Birke

A Century of Housing Struggles: From the 1915 Rent Strikes to Contemporary Housing Activisms. Edited by Neil Gray

Renewing Destruction: Wind Energy Development, Conflict and Resistance in a Latin American Context. Alexander Dunlap

Renewing Destruction

Wind Energy Development, Conflict and Resistance in a Latin American Context

Alexander Dunlap

London • New York

THE FOLLOWING RESEARCH CRITICAL OF RENEWABLE ENERGY IN GENERAL, AND WIND ENERGY IN PARTICULAR, SHOULD NOT BE USED TO SUPPORT, HELP OR ADVANCE THE USE OR CAUSE OF NUCLEAR, COAL, HYDRAULIC FRACTURING OR OTHER FOSSIL FUEL EXTRACTIVISM. On the contrary, the perspective advanced in this book seeks to cause deep critical reflection on the industrial system and energetic grid itself, promoting thought, imagination and experimentation to advance radical alternatives to development, 'progress' and modernity as we know it. The intention is to promote projects in the direction of enriching the quality of soil, water, food and human and non-human relationships to create long-term subjective well-being (happiness), individual/communal autonomy and ecological sustainability.

Published by Rowman & Littlefield International Ltd
6 Tinworth Street, London, SE11 5AL
www.rowmaninternational.com

Rowman & Littlefield International Ltd. is an affiliate of Rowman & Littlefield
4501 Forbes Boulevard, Suite 200, Lanham, Maryland 20706, USA
With additional offices in Boulder, New York, Toronto (Canada), and Plymouth (UK)
www.rowman.com

Copyright © 2019 by Alexander Dunlap

All rights reserved. No part of this book may be reproduced in any form or by any electronic or mechanical means, including information storage and retrieval systems, without written permission from the publisher, except by a reviewer who may quote passages in a review.

British Library Cataloguing in Publication Data
A catalogue record for this book is available from the British Library

ISBN: HB 978-1-7866-1065-2
      PB 978-1-7866-1066-9

Library of Congress Cataloging-in-Publication Data

Names: Dunlap, Alexander, author.
Title: Renewing destruction : wind energy development, conflict and resistance in a Latin American context / Alexander Dunlap.
Description: Lanham : Rowman & Littlefield International, [2019] | Series: Transforming capitalism | Includes bibliographical references and index.
Identifiers: LCCN 2019002454 | ISBN 9781786610652 (cloth : alk. paper) | ISBN 9781786610669 (pbk : alk. paper) | ISBN 9781786610676 (electronic)
Subjects: LCSH: Wind power—Latin America. | Power resources—Latin America. | Economic development projects—Latin America.
Classification: LCC TJ820 .D86 2019 | DDC 333.9/2098—dc23
LC record available at https://lccn.loc.gov/2019002454

This book is dedicated to all the communities and individuals who do not compromise: crossing into the unknown to advance their sense of autonomy and dignity in the face of immense forces and pressures to accept the political and economic orders that condition their lives and ecosystems.

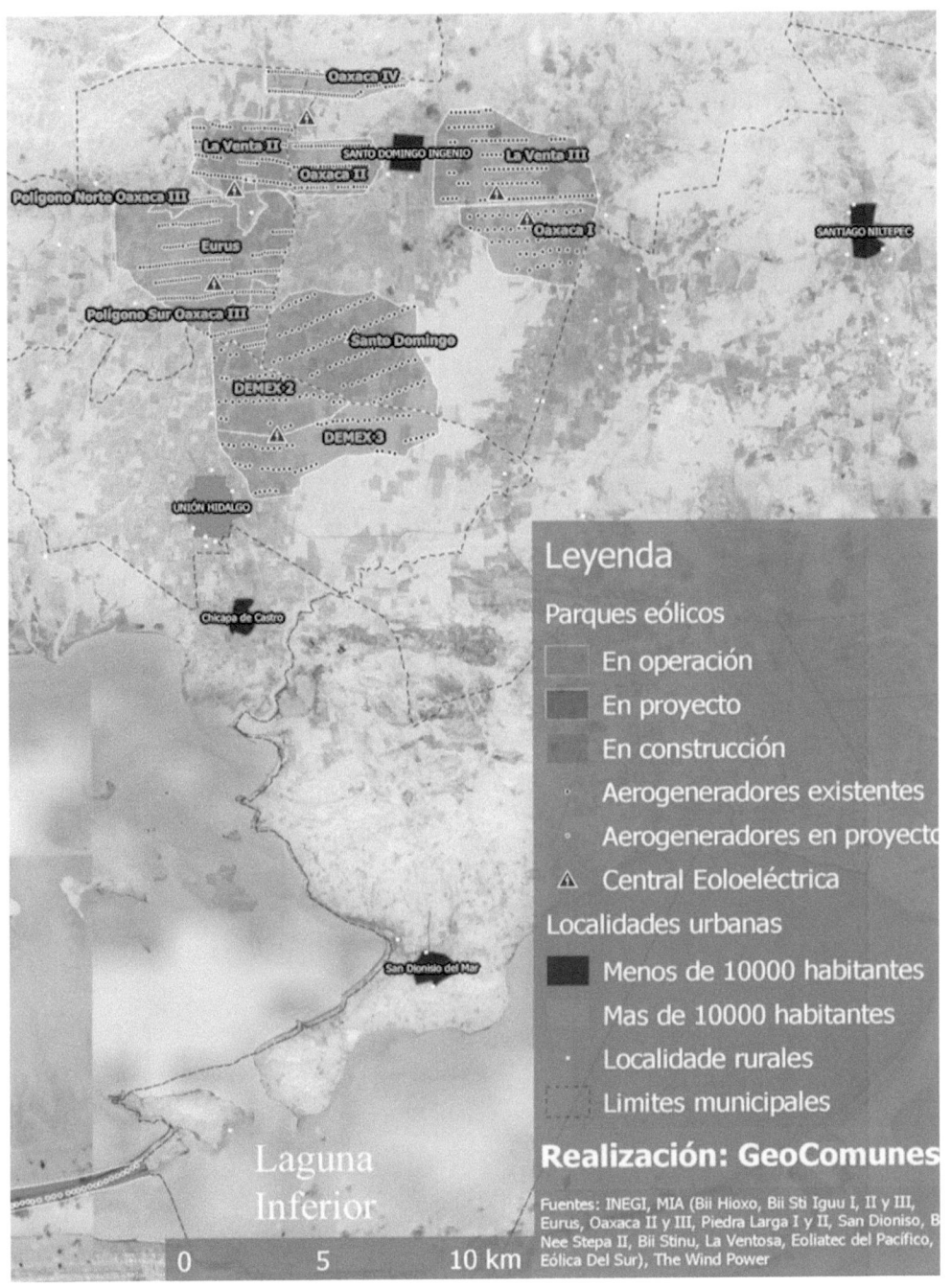

SOURCE: GEOCOMMUNES, HTTP://GEOCOMMUNES.ORG.

Contents

List of Tables and Figures. xiii

Abbreviations and Acronyms xv

Acknowledgements xvii

Prologue xxi

Introduction 1
The Journey into the Istmo   4
Discussing Anthropology: Positionality, Knowledge and Militancy   9
   For Anthropologists against Anthropology   10
Anthropology in the Field: Militancy   15
Literature Review and Theoretical Approach   16
Methodology and Research Approach   21
Book Structure   23

**Chapter 1: Welcome to the Istmo: A Brief History of Politics,
Conflict and Development** 29
From Conquest to Wind Turbines: Colonial Incursion,
   Adaptation and Resistance   29
   The Mexican State: Insurrection and Caciquismo   33
   The COCEI   37
   Wind Turbines   41
Conclusion   46

Chapter 2: 'We are surrounded': Living under Wind Turbines in La Ventosa 49
Wind Parks: Construction, Environmental Impact and Finance   50
   Cost, Finance and Greening Industrial Development   50

The Politics of Land Access and the Arrival of Wind Energy   53
   From Resistance to Negotiations: Land Contracts   58
Accounts of Environmental Impact   61
Land Change: Inequality, Rural Gentrification and Out-migration   65
   Temporary Work   66
   Electricity   66
   Rural Gentrification   67
   Crime and Drug Consumption   69
   Out-migration   70
Conclusion: Gains, Losses and Strangulation   71

Chapter 3: Counterinsurgency for Wind Energy: The Bíi Hioxo Wind Park 75
Counterinsurgency, Social Property and Indigenous People in Southern
   Mexico   78
Counterinsurgency in Mexico   79
From Death Squads to Arriving Wind Turbines   83
Divide and Conquer: Counterinsurgency for Wind Energy   88
   Hard Techniques   88
   'Soft' Techniques   91
Bíi Hioxo Realized   96
Conclusion   98

Chapter 4: Insurrection for Land, Sea and Generational Integrity in Álvaro Obregón 103
Zapotec Struggle: Gui'Xhi' Ro'   105
Wind Company Penetration in Álvaro Obregón   107
'We are the sea': Battle for the Barra and Municipality Takeover   110
The Constitutionalists—the Contra   116
The Communitarian Struggle Continues   118
Conclusion   123

Chapter 5: The Theatrics and Violence of Consultations: The Free, Prior and Informed Consent (FPIC) Consultation in Juchitán 127
Free, Prior and Informed Consent (FPIC)   129
Learning from the Past: From Barricades to Consultation   133
Free? Intimidation, Violence and Employment Opportunity   136
Prior and Informed   137
The Theatrics of Legitimizing Development   141
Conclusion   144

Chapter 6: Renewing Destruction: Colonization, the Genocide-Ecocide Nexus and Wind Energy Development 147
Welcome to Hell: The Colony Model   149
Colonial Genocide: Revisiting the Genocide Machine   152
Cultural Genocide and Wind Turbines   156
   Cultural Change: The Northern Coastal Istmo   158
   The Southern Coastal Istmo   161
   Wind Energy and the Genocide-Ecocide Nexus   163
Conclusion   165

Conclusion: The Grid System Spreads, Dependency Consolidates 169
Rebranding Dystopia and Rebellious Complicity   172
   1. Rebranding Dystopia   172
   2. Rebellious Complicity   176

Bibliography 179
Index 203

List of Table and Figures

Table 1.1.	Wind Parks in the Isthmus of Tehuantepec by Year. *Source: Author*	45
Figure P.1.	Solidarity Caravan Flyer in Viento Libertario, Issue 2	xxvii
Figure P.2.	Demonstration marches down Jóse Vasconcelos Boulevard.	xxvii
Figure P.3.	People hand banner on the walls of the military base.	xxviii
Figure P.4.	Demonstration heading to the prison.	xxix
Figure I.1.	Part of Álvaro Obregón's central gazebo mural.	8
Figure 2.1.	La Ventosa, Mexico.	53
Figure 2.2.	Mural in Juchitán: 'Autonomy and freedom of expression—My totopo will not have genetic modifications—Cacique oppress you—Political parties divide you—Enterprises invade us.'	69
Figure 3.1.	Security camera along the communal road in the Bíi Hioxo wind park.	87
Figure 3.2.	Photo from the Winds of Change Exhibition	95
Figure 4.1.	The Communitarian Headquarters, 2013: Veredas Autónomas	111
Figure 4.2.	Communitarian Police leaving for patrol	113
Figure 4.3.	Town Square Mural: 'Freedom is not conquered on your knees, but standing on your feet. Giving back hit by hit, inflicting wound by wound, death by death, humiliation by humiliation, punishment by punishment. Let the blood flow in streams because that is the price of freedom.'	123
Figure 5.1.	5 February 2015: Consultation House of Culture, Technical Committee during Q&A	129
Figure C.1.	The U-Form and M-Form corporate models.	173

Abbreviations and Acronyms

AEI—State Agency of Investigations
APIITDTT—Assembly of the Indigenous Peoples of the Tehuantepec Isthmus in Defense of Land and Territory
APPJ— Peoples' Popular Assembly of Juchitecos
CDI—Commission for the Development of Indigenous Peoples
CDM—Clean Development Mechanism
CDHT—Tepeyac Center for Human Rights
CERs—Certified Emissions Reduction Credits
CFE—Federal Electricity Commission
CTF—Clean Technology Fund
COCEI—Isthmus Coalition of Workers, Peasants and Students
COFEPRIS—Federal Commission for the Protection against Sanitary Risk
CONAFOR—National Forestry Commission
DAAC—Department of Land Affairs and Colonization
DGAC—Directorate General of Civil Aeronautic
EIA—Environmental Impact Assessment
FDI—Foreign Direct Investment
FN—Field Note
FPIC—Free, Prior and Informed Consent (consultation)
GLCC—General Law on Climate Change (2012)
GNF—Gas Natural Fenosa
IEEPO—State Public Education Institute of Oaxaca
IFC—International Finance Corporation
ILO—United Nation's International Labour Organization
IMF—International Monetary Fund
INAH—National Institute of Anthropology and History
IPP—Independent Power Production
IRENA—International Renewable Energy Agency
MW—Megawatt
NAFTA—North American Free Trade Agreement
PES—Payment for Ecosystem Services
PRD—Party of the Democratic Revolution

PRI—Institutional Revolutionary Party
PT—Labour Party
PROCEDE—Program for the Certification of Ejido Land Rights and the tiling of Urban Housing Plots
REDD—Reduce emissions from deforestation and forest degradation, and forest conservation, sustainable management of forests and enhancement of forest carbon stocks
SCT—Secretariat of Communications and Transportation
SEDESOL—Secretary of Social Development
SEMARNAT—Secretariat of Environment and Natural Resources
SME—Electrical Workers Union (Mexico)
SENER—Mexican Secretary of Energy
WDM—World Development Movement
WHO—World Health Organization
WTO—World Trade Organization
UCIZONI—Association of Indigenous Communities in the Northern Zone of the Isthmus
UN—United Nations
UNAM—National Autonomous University of Mexico
USAID—United States Agency for International Development

Acknowledgements

This book could not exist without the support and help of so many people from many different walks of life. While I often question my choice to attend graduate school and to continue on with the research presented here, none of this would have been possible without Aunt Beth. Everyone who knows her is well aware of her heart of gold, not only for her work with animals, but also for supporting me and encouraging me to go to graduate school. This support came when my life choices at that moment were either a bar or construction job; and at that moment I was still naïve about academic life and had an intense desire to read as many books as possible. This is a choice and possibility Aunt Beth provided me that I would not have had on my own, and for this I hold a great amount of gratitude and appreciation for her.

David Dunlap, who despite his madness, has always believed in me and supported me unconditionally no matter what, which is a parental strength that should never be underestimated. His support has been crucial to my development as an individual and is combined with support from others. Tony Puzynski, the best friend anyone could ever ask for, who has always been there to peel me off the ground when the going gets tough. Professor John Hall, who fulfills the archetypal supportive and inspirational professor, who has done well to inspire his students and stimulate their intellectual development, of which I am one of many. Professor Hall's work even extends to taking great risk to defend academic freedom in his classrooms from the formal and informal apparatus of repression deployed in classrooms to spy on students and professors alike (see Price 2005). Without professors like John Hall the university would be near worthless, regimenting people to fulfill roles as technicians and administrators of the everyday machinations of drudgery and terror (at its various levels, times and places) associated with modern life. Luckily, John is not alone. There are many professors who want to promote free academic explorations, curiosities and ideas of students in the face of intense work pressures, university economic restructuring and repressive Orwellian-style policies and programs, such as the US Patriot Act, the Pat Roberts Intelligence Scholars Program (PRISP) and the UK Counter-terrorism and Security Act, 2015. To these teachers and professors opposing and not negotiating with these programs in attempts to sustain spaces of free academic inquiry, I give acknowledgement and thanks.

The list also includes a special thanks to all of my supervisors and academic friends, past and present. Professor James Fairhead, thank you for introducing me to critical agrarian studies, creating space for me to follow my academic interest, taking a risk to write with me and also helping me survive the university experience with all of the unsavory bureaucratic and micro-political treachery that it entails. Ceri Oeppen, thanks for saving me while I was potentially homeless in the rolling hills of Southern England, you turned my lemons of life into lemonade! I still cannot stress my appreciation, not only for giving me an ideal exit plan when life took a difficult direction, but also for reading my work, having patience with me and believing in me as lecturer. Dimitris Dalakoglou I am torn with emotions. On one hand you are the most unreliable thesis reader I have ever had, while on the other hand you have an ethic of solidarity and care that is truly admirable. James, Ceri and Dimitris, thank for listening to me when I was in a rough spot in Mexico, respecting my assessment of the situation (as I was the one living there) and for not projecting your own worries and concerns onto me when I had to negotiate a stressful and uncomfortable situation—a time that with more stress could have easily led to my death, before anything else. You are all sparkling gems mined from the earth and placed in the banal stone and mortar fortification of the academy, which despite the histories of pillaging Africa, Latin America and 'impoverished' rural areas of the world, the university is less an ivory tower and more a dilapidated medieval castle.

Before moving into the beautiful and courageous people who I lived and worked with in Oaxaca, there are two other professorial gems worth mentioning: Raminder Kaur and Ton Salman. Raminder, thanks so much for your patience, conversations and commenting on my work, it was not only helpful to my academic development, but enjoyable. Not to mention your joyful spirit and realistic assessment of academic micro-politics. Our reading group and conversations were engaging and enjoyable. This also includes appreciation for the Journal of Economic Issues (JEI) and the Association for Evolutionary Economics (AFEE) for allowing permission to reproduce William Dugger's diagram of corporate technological restructuring (see Figure C.1). And Ton Salman you were very important to this book. Our time together has been short and my disposition needy, but that bright smile of yours has held out over the years of academic life, which is not only impressive, but truly reflects the greatness of your character and commitment as a reliable supervisor to answer questions, read and comment on my doctoral thesis. Your comments, Ton, have proved invaluable and yours skill to help guide writing dead-ends is of the most impressive I have ever experienced or seen. Finally, before moving on to Oaxaca, I would like to thank the editors of the *Transforming Capitalism* book series and the staff at Rowman & Littlefield who worked on this book (Dhara Snowden, Rebecca Anastasi and other nameless individuals). This includes a special acknowledgement of Ian Bruff for their patient reading and thoughtful copyediting of this book. My appreciation, gratitude and admiration goes out to all of you, and if there was anything good that came from academia it was my conversations and experiences with all of you.

This leads to all the people who helped with this research and worked to keep me alive in Oaxaca. Professor Corbett, thank you for overseeing and taking an interest in my research. It must be added that it has been a real privilege to know Jack, be subjected to his storytelling and his teaching through puns, and his honorable commitment to academic freedom. This leads to love, care and appreciation for people who were fundamentally important to my safety, fun and research who will only be mentioned by first name as the situation in Oaxaca is treacherous and is becoming more so by the day. My Spanish school, you know who you are! Anabel thanks for introducing me to your friends and family, I look to forward to crossing paths with you in the not too distant future. Ernesto and Gloria, not only do I appreciate your taste in music, but your taste in food. You two are amazing friends. I am so happy to have met and spent time with the two of you and look forward to seeing you in the future. Nadia and Anna, another wow! You two were my rocks and your support in Oaxaca was foundational to helping me live comfortably and survive rough times, which I recognize is normal life for so many in Oaxaca. Thank you for sharing your homes with me and my friends. Also Paul R., I appreciate your commitment to translate with me in my outdoors office against the backdrop of your intense and busy schedule—thank you. Ivan, *la banda*! I send my love to you and your family, thanks for looking after me and I hope your son's skills as a guitar player are continuing to bloom. I hope if any of you read this it finds you all happy, healthy and enjoying life.

To the people active in the struggle, all too aware of the social war that engulfs their lives; I want to express my gratitude for your disposition and support. First, Mr. X—*odelay*—I hope you are still alive and are happy running in the streets with your gang of street dogs. I want to apologize formally for how difficult I can be and what a workaholic I am—for that matter this extends to anyone and everyone who has worked closely with me—I can be a difficult person, and I most certainly was, and you know that better than anyone. Mr. X, I send a deep gratitude for your patience and all the work you did to make this project possible—without you I would not have been able to go as far as I did in such a short period of time. Thanks for being you, with all the difficulty that entails, and I can only hope that if and when you read this book, a smile of satisfaction can come from your face—even if none of it was easy and all of it was dangerous. Flaco thank you for enthusiastically supporting this project, your ideas, experience and our conversations have undoubtedly influenced this book, and like Mr. X, I hope you are still alive and happy.

This extends also to all the communities in resistance and struggling against wind energy development that embraced me and helped me to get an understanding of what was happening in the region. Let a grand thank you be extended to those fighting against wind energy exploitation. Radio Totopo, the *Asamblea Popular del Pueblo Juchiteco* (APPJ) and the Assembly of the Indigenous Peoples of the Tehuantepec Isthmus in Defense of Land and Territory (APIITDTT), thank you for trusting and working with me when the odds were stacked against me. Your support, care and concern were invaluable to me, and I hope this book will prove valuable and helpful in your struggle for existence and justice against wind energy development and other oppressive forces in the Istmo. The *cabildo comunitario* in Álvaro Obregón, especially select members who stood by me in difficult times of repression, thank you for trusting me. *Los Vatos Locos* of the region fighting in defense of their land and territory, I enjoyed my time with you and I hope this book finds you in a good place, ideally with coconut mescal and *caldo de pescado*. Without all of you, your support and trust, none of this could have happened, and I hope this book can serve as a projectile in your struggle, even if the pages of this book can only serve the purpose of rolling papers in your struggle.

With gratitude and appreciation.

Sincerely,
Alexander Dunlap

Prologue

Two weeks before arriving in Mexico to research wind energy projects in the Isthmus of Tehuantepec region of Oaxaca, and just over two months after the *Ayotzinapa*, forty-three student disappearances were carried out in Iguala, Guerrero (Paley 2015), I sat through an academic panel at the University of Sussex organized to discuss the disappearances. On 26 September 2014, six people were killed, three of them were students, one of whom had his face peeled off and yanked around his neck, while another forty-three disappeared in a joint effort by the police, military and Narcotics traffickers (Paley 2015). This panel was disappointing to say the least, just rehashing the same narratives about Mexico as a failed state, and all of the other cliché approaches to (online) 'activism' that academia can offer. One could read the panel as doing everything to make Mexico appear as an isolated exception—a 'bad apple' within the state system—forgetting how deeply integrated Mexico is in global business, finance and even the geopolitical strategies of the United States, Canada and to a lesser degree Europe. While only mentioning in passing the enormous amount of counterinsurgency aid from the United States,[1] the panel failed to discuss at any length the material support in technology, weapons, and training for the Drug War in Mexico, that in practice articulates itself as a low-intensity war against the entire population targeting students, organized labor and Indigenous communities (Paley 2014; AI 2015; Correa-Cabrera 2017). The academic panel superficially addressed the changes taking place in Mexico under the guise of the War on Drugs under the Mérida Initiative, or Plan Mexico. The restructuring and expansion of the Mexican courts, prison system and military/police apparatuses were not mentioned; nor were the appearance of US intelligence fusion centres (Paley 2014, 117). What was discussed was how these forced disappearances were getting significantly more media attention than others, which was odd as this kind of state violence is all too frequent in Guerrero, Oaxaca, Chiapas and other states in Mexico where military, police and extra-judicial forces are well-known for committing atrocious human rights violations with impunity (Gibler 2009; HRW 2009; Norget 2005; Stephen 2000; 2002; 2013).

There have been 40,180 people disappeared, of which around fifty percent occurred between the years 2012–2014 (AI 2015). Since the beginning of the Mérida Initiative in 2007, there have been upwards of 150,000 people murdered, fifty journalists killed, and the widespread use of kidnapping, and the worse kinds of torture employed by both state and non-state actors in Mexico (AI 2014; Paley 2014; Correa-Cabrera 2017). Additionally, these acts of state violence occur alongside the generalized attacks against migrants by both state and non-state actors, where they are robbed, raped and murdered as they travel through Mexico from Central and South America to the United States (AI 2014; 2015).

These events take place against the backdrop of thirty years of economic restructuring and privatization. Mexico received thirteen structural adjustment loans from the World Bank between 1980–1991, while around the same years 989 state owned businesses were privatized, leading to changes in Article 27 in 1992 that created the possibility of buying, selling and privatizing social property—*ejidos* and communal land. This coalesced into the well-known North American Free-trade Agreement (NAFTA) in 1994 that opened up the country to foreign companies and served as the spark launching the Zapatista uprising (Stephen 2002; Bello 2004). This trend has been forcefully continued to the present (Carlsen 2008; Paley 2014; 2015; Correa-Cabrera 2017) with two significant privatization measures put forward by Mexican President Enrique Peña Nieto in 2013. The first was his education legislation designed to weaken the teachers unions and lay the foundation to privatize public education in Mexico (BBC 2013). This can be read as reprisal for the social activism of teachers unions in general, but specifically the politically engaged Oaxaca's Section 22 teachers union who annually strikes for better working conditions and in 2006 triggered a seven-month insurrection and commune against the authoritarian Institutional Revolutionary Party (PRI) Governor Ulises Ruiz Ortiz (see Gilbert 2009; Stephen 2013; Jenss 2018). The new education legislation has caused widespread outrage and protest that has only escalated with militant protests and deadly repression in Oaxaca City, nearly igniting another insurrection ten years later in June 2016 (Campbell 2016). Equally controversial is the Petroleum Act and Federal Electric Utility Act passed on 21 December 2013 which came into effect in the summer of 2014 (Payan and Correa-Cabrera 2014; Correa-Cabrera 2017). The struggle against energy structural adjustment has been going on for decades as it privatized the seventy-five-year-old Mexican oil company PEMEX as well as Central Light and Power (LyFC) and the Federal Electricity Commission (CFE) (Cypher 2014). This was done based on claims that the national oil company's public status was making the industry stagnate, notwithstanding it was one of the only companies with profits, in 2004, that were $40 billion, while the ministry of finance taxed the company $42 billion, leading critics to think this was an intentional attempt to bankrupt PEMEX and force it into the private sector (Navarro 2013). This legislation is advancing the privatization of social property, ceding immense power to resource extraction companies, attempting to make it obligatory for landowners to negotiate their land at a 'fair market value' to energy companies which in theory makes it so 'refusal is not an option' (Payan and Correa-Cabrera 2014, 4).

Economic structural adjustment laws and state violence is deeply intertwined (Paley 2014; Blakeley 2009; Bello 2009; HRW 2009; Klein 2007; Stephen 2002; 2000) and, tragically, the massacre of *Ayotzinapas*' forty-three students caused little reflection within the Mexican government, who publically unfazed, would use these events to further an agenda of Economic Liberalization. 'Regardless, the tragic events in Iguala also reveal a social and economic dimension,' says President Peña Nieto who did not waste time announcing in November that:

> Today there are two Mexico's: One, inserted in the global economy with increasing income levels, development and well-being. And on the other hand, there is a poorer Mexico with age-old backwardness that has not been resolved for generations. . . .
> I propose the establishment of three special economic zones *in the most backward regions of the country*. These are: The Inter-Ocean Industrial Corridor in the Isthmus of Tehuantepec, which will connect the Pacific with the Gulf of Mexico; the second, in Puerto Chiapas, and the third, in the municipalities that are connected to the Port of Lazaro Cardenas in Michoacán and Guerrero. A special economic zone is an area where we will offer a regulatory framework and special incentives to attract corporations and generate quality employment (Quoted in Paley 2015, 6, emphasis added).

For the Mexican government, deregulation, foreign direct investment (FDI) and corporate control has become the 'be all and end all' solution to social problems in Mexico—even state sanctioned student disappearances. This statement by the president not only condones acts of state violence, but also the previous failures of government policy as well as writing off the diversity of people in these regions and their cultures as 'backwards', demonstrating an unapologetic and colonial mindset for crimes committed against the population. In this moment, Mexico embodies a duality. On the one hand, the Mexican government orchestrates a Dirty War with increasing and widespread securitization polices, resulting in disappearances, massacres and human rights violations with impunity (see Paley 2014; HRW 2009; Norget 2005; Stephen 2000; 2002; 2013; Correa-Cabrera 2017). On the other hand, it appears that Mexico is articulating some the most progressive environmental legislation, promoting the development of a green economy, signing ambitious climate change legislation and working towards renewable energy transition. This raises curiosity about the relationship between these two tensions of abhorrent state-narco violence and seemingly progressive environmental policy. Examining where these two trends meet and what exactly is the outcome of climate change mitigation initiatives and shifts towards a green economy in Mexico are the principal questions underlining this research on wind energy development in Oaxaca, Mexico.

ARRIVING IN MEXICO

After travelling twenty hours from London on 12 December 2014, I arrived in Oaxaca City. My reasons for traveling to Oaxaca were to conduct research on the social impact of industrial-scale wind turbines in the Isthmus of Tehuantepec region—known locally as the *Istmo*. Addressing the coastal *Istmo* in two distinct regions—the north and the south—will be helpful in conceptualizing the myriad changes in the conflict. The northern part of the coastal *Istmo* had experienced intense wind energy development since 2003 when wind companies began planning projects in the mid-2000s in the south, later attempting construction around the Laguna and Pacific Ocean in 2011. While there have been regional negotiations, disagreements and conflicts for ten years around these projects, it was not until they started moving into the territories of subsistence fishing communities around the Laguna that the conflict against wind energy began to spread and generalize. On the ground, assemblies in towns surrounding the Laguna in the *Istmo*—San Dionisio del Mar, Álvaro Obregón, San Mateo del Mar, San Francisco del Mar, Santa Maria Xadaní, and Juchitán—have been resisting the spread of wind energy projects for several years, which in Álvaro Obregón, or Gui'Xhi' Ro' in Zapotec, has now extended to rejecting all political parties and rejecting elections imposed by Federal and State authorities. The resistance against wind energy, in all its diversity, tends to view politicians as self-interested proxies acting on behalf of the wind companies, while elections act as a mechanism to legitimize and facilitate the takeover of the land and sea (APIIDTT 10 Dec. 2014). Other opposition groups on the other hand, work with electoral figures to various degrees; while politicians, select land owners and other collaborators view these wind projects as a source of revenue, work and a step towards progress and development.

Just a week before arriving in Oaxaca City, resistance had manifested itself in San Dionisio del Mar with a road blockade to monitor and prevent wind companies, politicians and members of the National Electoral Institute (INE) from entering the town—the latter requested by the local Institutional Revolutionary Party (PRI), which was actively collaborating with the wind companies. On 5 December 2014 two people were injured after a confrontation at this blockade, while on 14 December the conflict escalated involving gunfire, sticks and stones as six people were injured including a pregnant woman (SIPAZEN Dec. 2014). This triggered demonstrations throughout the town. The People's Assembly of San Dionisio del Mar (ADPSM) denounced these provocations on 16 December, stating:

> given that it was anticipated that the conditions do not exist to carry out municipal elections due to the *sequelae* [consequences of a previous circumstances] left by our recent struggle against the wind-energy firm Mareña Renewables, and in light of the post-electoral conflict which consequently degraded the social fabric [of the town], the federal and state governments seek to install functionaries in the municipality so as to open the door to these [wind energy] firms (SIPAZEN December 2014).

Six days later the situation was continuing to escalate. Five hundred state police and State Agency of Investigations (AEI)[2] officers laid siege, temporarily occupying San Dionisio del Mar in an attempt to re-establish law and order. This attack resulted in intense fighting on the barricades, with one police car being torched on the access road where the community had established their checkpoint. Eventually, the police broke through and occupied the town, forming a caravan of vehicles that some would later call 'the black snake' that drove in and along the outskirts of the town. This police occupation ended a few days later and the election was suspended without a new election date—a temporary win for the resistance (SIPAZEN Dec. 2014).[3] This was the context in the *Istmo* when I arrived in Oaxaca City and began preparing to move to the region to investigate the social impact of wind energy development.

While the conflict was fomenting in the *Istmo* over wind turbines, in Oaxaca City, I attended Spanish language school and was preparing for my research. It was the holiday season and the streets of Oaxaca were overflowing with bright-eyed tourists from different parts of Mexico and the world— shopping, partying and celebrating the holidays. This year, however, the tourists were greeted by theatrical performances, soap box speeches and an occupation in the Zócalo at the center of the city, all of which were denouncing the Mexican government's state violence, collaboration with Narcos and its failure to properly investigate the disappearances in Guerrero state. This led the Attorney General of Mexico during a question and answer session about the Iguala disappearances to end the session by saying: 'Enough, I'm tired'—putting an end to public dialogue (Pizarro 2014). Tourism roared on against a backdrop of an occupation and an entire city covered with graffiti saying: '43,' '43,' '*Faltan 43*' (The missing 43), '*Nos Faltan Los 43 Maestros*' (We are missing 43 teachers), and '*Ayotzi vivos, Vive Los 43*' (Ayotzi lives, the 43 live). A stencil on the wall caught my eye that depicted the president of Mexico as the Grim Reaper saying: '*Asesino. Peña Nieto Asesino Faltan 43*' (Killer. Peña Nieto is the killer of the Missing 43). Nevertheless, tourism continued apace with walls and people in protest, but the booming and bustling of tourist shopping, eating and cultural consumption would soon meet with an influx of police.

A little over a week before and after Christmas, a full-blown police occupation would engulf the city centre. When I asked locals about the large police presence, I was told that it was to discourage the robbing of tourists, a concern related to the *Zócalo* occupation. Despite the widespread executions, disappearances and torture of civil dissent, the tourism industry seemed unaffected, giving the impression that Mexico's Dirty War could co-exist or was even complementary with tourism and economic growth.

When the holiday season started to come to a close, anxiety about beginning my research was starting to take hold. I had assumed that I would try to live in San Dionisio del Mar, but I still wondered if it was *the best* research location. The next step soon presented itself. I heard from friends that a solidarity caravan was organized to support the people struggling against wind energy development in the *Istmo*

I was excited. Not only would I be able to go to the area, but I would do so with people in support of the resistance. Uneasy because I had little information about this caravan (it sounded too good to be true), and all I knew was a date: 15 January. One night, a couple of days before the date, I went to a local bar. I had received information about the caravan including that I needed to write an email to confirm my seat. I had done so, but two days passed, yet I never received a reply. On the door of the bar, I saw a poster for a movie about the wind energy conflict in the *Istmo* by Alèssi Dell 'Umbria. The film was titled, *Istmeño: Viento de Rebeldia*. This sparked enthusiasm, leaving no question about my attendance at the event.

The days passed and friends and I went to the film venue. It was the kind of Oaxacan bar that is designed and decorated for tourist pleasure, providing a heavy dose of Oaxacan style in a sanitized and comfortable environment for foreigners. The restaurant walked a careful line between traditional Oaxaca and corporate culture that made for an enchanting fusion for those who could afford it. As I waited for the film to start in a crowd of at least fifty people, I saw that a newspaper was being handed out. Catching my attention, I rushed for a copy. It was an anarchist paper entitled: *El Viento Libertario*.[4] The cover displayed an image I recognized from the blog of the Indigenous Peoples' Assembly of the Isthmus in Defense of the Land and Territory (APIITDTT).[5] It was the logo of the Communitarian Police (*polícia comunitaria*) of Álvaro Obregón/Gui'Xhi' Ro'. Here, however, several anarchist circle-A's were clustered under the logo. This immediately signaled that there was some type of anarchist support and/or presence in the area of my research, sparking increasing interest on my part.

Grasping the paper, I sat through the extensive two hour documentary that provided a direct and comprehensive voice to the people in the area. After the film, three people from the towns in resistance addressed the audience. I recognized them from an earlier documentary about their struggle,[6] as they explained how troubling the conflict was while relating their words to the issues depicted in the film. These speakers were followed by the three anarchists who had distributed the paper, presenting to the crowd an invitation for the Libertarian Caravan. 'Aha!' I thought, 'these are the people I have been emailing without reply!' I approached them and told them I was a doctoral student in anthropology researching the social impact of wind energy on people in the *Istmo* and that I wanted to be on the caravan (see figure P.1). After a short conversation, they gave me information and told me that the caravan would leave after a demonstration for the disappeared forty-three teachers (*normalistas*). The demonstration would start at the Zapata statue at 10.00 the next morning. The puzzle pieces for this research project were slowly coming together. Pleased to be joining the demonstration condemning state disappearances, and even more excited about the solidarity caravan, I returned home and began to prepare for my second journey to the *Istmo*. This time, however, I would visit the areas in resistance against these large-scale wind energy projects.

I arrived at the Zapata stature with my backpack and sleeping bag in hand. Arriving early, I watched as people filtered into the area about an hour after the meet-up time. Eventually thirty-five people grouped around the statue with their backpacks and stencils in hand.

Slowly people started to move and change their clothes, putting on masks, taking over the street and pulling out a big banner to stretch out across the entire road. The banner read '*La Vía Electoral Fracaso . . . Y Siempre Fracasara. ORGANIZACIÓN—*

Autonoma Autogestiva' (The electoral way failed, and will always fail. Autonomous and self-managed organization; see figure P.2). We took to the streets shouting slogans against the state capital and their prisons and police, while demanding the return of the disappeared: '*Vivos se los llevaron, vivos los queremos!*' (They took them alive, we want them back alive!). The march went up Eduardo Vascocelos Blvd. Stencils were spray-painted on vacant walls and corporate property. Traffic was brought to a pedestrian pace. Walking along Vasconcelos Blvd., people were

Figure P.1. Solidarity Caravan Flyer in Viento Libertario, Issue 2

Figure P.2. Demonstration marches down José Vasconcelos Boulevard.
SOURCE: NOTICIAS

shouting, spray-painting slogans and wheat-pasting posters, meanwhile motorcycles with two soldiers on each followed the march with AR-15 assault rifles slung over their shoulders. They kept their distance as the demonstration continued down the highway without any altercations. At an intersection, I saw a large road sign pointing to the *Istmo* (Isthmus of Tehuantepec region, see figure P.4). Looking at the sign, I realized that in the context of the widespread state terror being exercised against the population, my fieldwork would be carried out in protest. The demonstration continued down the highway and turned on to Chapultepec Road, heading for the Central State Penitentiary Ixcotel that is connected to the 28th Military base.

By this time reporters were surrounding the demonstration and taking photos.[7] At the entrance of the military base, soldiers stood guard in bunkers and guard towers with assault rifles and other armaments. Fearlessly, demonstrators spray-painted messages of protest and discontent on the walls (see figure P.3), while others stood in front of the soldiers holding banners. I was shocked at how demonstrators asserted themselves in the face of these heavily armed soldiers.

People continued to shout and yell denunciations while spreading their message of disapproval, condemning paramilitary violence, political repression as well as the student disappearances. Slowly this march and its lines of paint continued along the side of the military base up to the entrance of the prison, where people became more confrontational. Masked women threw rocks at the police trucks parked behind the entrance checkpoint. Time passed and denunciations were read to the media,[8] while people continued chanting slogans and a disagreement surfaced with visiting families of prisoners. The military and police did not attack the demonstration and

Figure P.3. People hand banner on the walls of the military base.
SOURCE: MENDOZA, 2015

Figure P.4. Demonstration heading to the prison. SOURCE: AUTHOR

after a couple hours people started to leave in large groups, filtering into the streams of commuter traffic. Many of them would continue on the solidarity caravan leaving for the *Istmo* that evening. Given how this research started in protest, it might not be surprising that five months later, I would leave the *Istmo* early because local authorities working with the wind energy companies hired a gunman to target me.

NOTES

1. Total US spending 2008–2014: $2.35 billion with congress fighting for $115 million in 2015, which was not approved (Paley 2014, 87).
2. Mexico's Federal Bureau of Investigation (FBI).
3. This changed on 6 June 2016 when the first female mayor was elected in San Dionisio del Mar (Marzo June 2016).
4. Available here: https://vientolibertario.files.wordpress.com/2015/01/viento2.pdf.
5. Available here: https://tierrayterritorio.wordpress.com/.
6. Available here: https://www.youtube.com/watch?v=PZ35UO_J3og.
7. Available here: http://www.noticiasnet.mx/portal/oaxaca/general/protestas/256867-protestan-anarcos-cuartel-militar-penitenciaria-oaxaca.
8. Available here: https://www.youtube.com/watch?v=bN8AsbvYdHQ.

Introduction

This book examines the conflicts and complications generated by wind energy development in the Isthmus of Tehuantepec region of Oaxaca, Mexico. Known locally as the *Istmo*, this region is located in the southwest corner of Oaxaca state and has been subject to ten years of intense wind energy development, which has no end in sight. The arrival of wind parks in this region are largely the result of a 2003 USAID report, *The Wind Energy Resource Atlas of Oaxaca* that established the *Istmo* as an 'excellent resource (power Class 7)' site for wind energy generation (Elliott et al. 2003, iv). Since then the International Finance Corporation (IFC 2014, 1) has declared the *Istmo* 'home to some of the best wind resources on earth'. The first wind project in the *Istmo* was a seven wind turbine pilot project developed by the Federal Electricity Commission (CFE) in 1994. It was not until 2003, however, that a 'wind rush' engulfed the region with transnational consortiums arriving to the *Istmo* to build and operate wind parks. By January 2015, 1,608 wind turbines had been built, mostly located in the northern part of the coastal *Istmo* (Rivas 2015). La Ventosa, a town in the northern coastal *Istmo*, has seen itself surrounded by wind turbines over the last six years (see Chapter 2). The Mexican government and wind companies have been working to spread wind energy parks across the entire region, notably to the southern areas around the Laguna Superior, the Barra de Santa Teresa and the Pacific Ocean. Wind energy is justified in the *Istmo* in the name of foreign direct investment (FDI), mitigating anthropogenic climate change and enacting strategies of sustainable development. As this book will demonstrate, the story of wind energy is more complicated than the government and wind companies would have us believe.

Planning for the Bíi Hioxo wind park began in 2006 on the outskirts of Juchitán de Zaragoza, comprising 117 wind turbines managed by Gas Natural Fenosa (GNF). This wind project was militantly resisted by primarily Zapotec (*Binnizá*) and Ikoot (*Huave*) farmers and fishermen living in the area, as they viewed these projects as a neocolonial takeover that, contrary to grandiose claims of bringing employment, social development and prosperity, wind projects brought social divisions, and destruction to the land, sea and what remains of Istmeño cultures. Despite the efforts of those resisting the Bíi Hioxo wind park, it was the first wind park constructed on the Laguna superior, in October 2014, a case discussed at length in Chapter 3. Meanwhile, the Mareña Renovables was trying to build 102 wind turbines on the sand bar that divides the Laguna Superior and Inferior called the Barra de Santa Teresa or *Barra* for short, in addition to thirty more turbines on the Pacific Ocean around San Mateo del Mar (Howe et al. 2015; Howe 2014; CDM 2012b; Smith 2012). This project was challenged by a variety of Indigenous communities in the region. San Dionisio del Mar, Álvaro Obregón and San Mateo del Mar, among others, engaged in militant resistance and a legal strategy against the project. On 7 December, 2012, the Seventh District Court in Salina Cruz officially halted the project with a court injunction (*amparo*) (Petersen 2012). People living around the Laguna told me that this court ruling has not stopped the wind companies from trying to enter and build on the Barra, which is discussed in Chapter 4. Mareña Renovables would later change its name to Eólica del Sur and relocate inland between the cities of Juchitán and La Ventosa. After ten years of wind energy development along with protest and resistance from communities, this new wind project would eventually be accompanied by a Free, Prior and Informed Consent (FPIC) consultation that is discussed in Chapter 5.

The drive to build wind parks only increased with international concerns about anthropogenic climate change. The promotion of wind energy has become an important element of green economic development and a cornerstone of climate change mitigation underscored by the United Nations Framework Convention on Climate Change (UNFCCC 2012, 22) to 'accelerate the deployment of all renewable technologies'. This has made the Isthmus of Tehuantepec an important and valuable national asset to Mexico, now a leader in promoting a green economy and climate change legislation with the Renewable Energy and Energetic Transition Law (2008), the Special Climate Change Program (2009–2012) and the General Law on Climate Change (2012) which seeks to reduce emissions by 2020 to thirty percent of year 2000 levels and fifty percent by 2050. This plan is known as the 10-20-40 vision, and according to Secretariat of Environment and Natural Resources (SEMARNAT 2013, 9), seeks to turn:

> this great [climate change] challenge into an opportunity to conserve and sustainably use its natural capital; to take advantage of its vast potential to develop clean energies; to correct inefficiencies in the use of energy; to generate jobs within a green economy; to promote sustainable territorial development; to increase competitiveness, and to improve public health and quality of life.

Mexico's countrywide trajectory started with the General Law of Climate Change (LGCC), which institutionally supports and mandates the expansion of renewable or 'clean' energy. According to the LGCC, this will be accomplished with 'an incentive-based system, which promotes and allows for profitable electricity generation through renewable energy such as wind, sun, and small hydro' that seeks to meet Mexico's goal to generate thirty-five percent of its electricity from clean sources by 2024 (LGCC 2012 65; SCCP 2014). The importance of developing an economic incentive structure and reducing friction against wind energy development in the *Istmo* is also outlined in the USAID (2009b) report "Elements for the Promotion of Wind Energy in Mexico".[1] These ambitious climate change laws and green economy policies are further embedded in the Special Climate Change Program 2014–2018.

While wind energy has the potential to address ecological degradation and climate change, this book demonstrates, however, there are structural problems associated with wind energy development tied to centuries-old patterns of grabbing Indigenous land, industrial development, and legally binding growth imperatives that mandate economic and material growth and (see Nace 2003), consequently, the increasing electricity demand from private sector industries. These issues are highlighted by Indigenous groups resisting wind energy on their land and territory, raising issues with industrial development, (neo)colonization and capitalism, especially as they manifest in widespread ecological catastrophe, anthropogenic climate change and the circumscribing of human and non-human diversity. Reflecting on insights from the previous chapters, the issue of genocide and ecocide in relation to wind energy development will be discussed in Chapter 6.

Although resistance against wind energy is real and cannot be ignored, it must be noted that some locals embrace and work with the wind companies. In fact, local elites, politicians and some land owners are actively promoting and fighting for wind park development. This has led to strife and divisions within villages and families that further complicate the micro-politics of the conflict and the politics of cultural change and/or destruction. By revealing the strategies and tactics devised to promote state stabilization and control through the lens of wind energy development, this book will thus secondarily contribute to an anthropology of the state (see Das and Poole 2004; Nuijten 2003).

The conclusion then takes the case study and theoretical development of the book further by arguing that wind energy development in the *Istmo*, or renewable energy in general, continues to consolidate, intensify and expand capitalist infrastructure and relationships, state violence and infrastructural development that, in its present form, is altering local livelihoods, cultures and ecosystems. If this present trajectory continues, it will lead to significant cultural and ecological degradation, if not destruction—hence the title of this book: *Renewing Destruction.* Wind energy, in its current and industrial-scale manifestation, is renewing the destruction of the industrial-capitalist system and not de-growing, transitioning or repairing socio-ecological damages brought by industrial development (as it is popularly envisioned by the public).

This book is an exploration of wind energy development in the *Istmo* and all the complications that might entail. The next section begins where the prologue left off: with my initial fieldwork encounter, the intense situations that I confronted and, consequently, how the research sites were chosen. I will then discuss the position and role of the anthropologist, while revealing my ethical approach and position during this study. Following a review of the existing literature on land-grabbing and wind turbines, I will discuss the theoretical approach underlying this research project before moving into the methodology used to research the impact of wind energy on communities in the *Istmo*. The Introduction will conclude by outlining the structure of this book.

THE JOURNEY INTO THE ISTMO

The anarchist demonstration in Oaxaca City protesting the *Ayotzinapa* murders and disappearances disbanded late in the afternoon of 15 January 2014, and by midnight I was on a bus heading to the *Istmo*. Because of the current state of Mexico and the political nature of the solidarity caravan, I was told that the buses were being monitored by human rights groups using GPS. This news accompanied an air of subtle anxiety about the possibility of an attack by security forces. In Mexico, the police, military, paramilitaries and Narcos are widely understood to be in collaboration.[2] Other than a traffic stop by a solider before leaving Oaxaca City, the caravan had no problems on its way to the *Istmo*. I continued to stare out the window, listening to the chatter of the highway before I eventually curled up across two bus seats and passed out.

I woke some hours later on the Pan-American Highway, towered over by columns and rows of enormous wind turbines as a twilight red sunrise set a dystopic tone. The notorious *Istmo* wind was howling so fiercely that at one point it blew off the emergency exit hatch on the roof of the bus. Most of us jumped and gasped and then shock quickly dissolved in a soft laughter. I could not help but think how cliché an introduction this was to the *Istmeño* wind, a wind well-known for blowing over semi-trucks and trees. In a sleepy daze, I was fascinated by finally seeing these fields of wind turbines that I had read and heard so much about and becoming more acquainted with the communities resisting the spread of these giant robots harnessing the vitality of the wind.

The first place we arrived was a communal primary and secondary school in the far south of the region in San Francisco Ixhuatan. This area was past all of the lands occupied by wind parks. After arriving, we ate breakfast and later gathered in a classroom to meet with the school's director who talked to us about the school and its relationship to the wind parks. The town was in Ikoot (*Huave*) territory and there were two schools in the town. We stayed in the school that taught and practiced *comunalidad* which is a form of education designed around indigenous cultures in Oaxaca; a life philosophy rooted in sharing every facet of life, and organized around territory, governance, labor and enjoyment (Luna 2010; Manzo 2011; see also Gross 2015). The director explained that they were developing a drama program, sports and a community radio program alongside other primary education curricula. Given the school's concern for community issues, they eventually took a public stance against the wind projects. The director explained:

They [the Mexican government] are angry because the school has launched an information and consciousness raising campaign with information about the wind energy projects on the Laguna Superior's shore, the effects of mining in the Chimalapas or about the salt mine on our beach in Aguachil, which is a sacred place for us.

This information campaign soon resulted in repression. First, directly with the theft of the community radio in the school and, secondly, indirectly with institutional attacks from within the state's educational institution IEEPO (Instituto Estatal de Educación Pública de Oaxaca) and the other school in town, which has an administration affiliated with the local Institutional Revolutionary Party (PRI), who are active supporters of the wind energy projects. The director continued:

One year ago somebody stole from us, which means we are not safe. It began when we started to show placards [against the wind energy projects] in parades and eight days later, somebody broke down the door and all the video equipment, cameras, microphones, projectors—everything [was gone]. It was the municipal government. After that we are on alert. [. . .] We have not been aggressive because the school is for pre-university people. We just showed our placards on parades in the community, but that led some people invite us for talks in town hall. [. . .] [S]o people see us like an anarchist school or ask what we are doing, because we are receiving kidnapping and death threats.

It was shocking to hear about this level of repression being leveled against an elementary school for taking a position against the wind energy projects. The director went on to describe how wind company representatives are 'buying consciences', which he said had instigated a rise in 'inter-communal conflicts'. Although this area had no wind turbines, it is one of the parts of the *Istmo* where future wind parks are proposed, and the school's philosophy and political expressions—making posters and marching—earned it repression.

After the talk, people took walks, swam in the river and talked together, and later performances were put on by members of the caravan. These included comical acts,³ arts and craft workshops with local kids and presentations about anarchist ideas and struggles outside the *Istmo*.

The next day, everyone woke up, ate breakfast and, I particularly, enjoyed how the school used neighbouring sheep to cut the grass. We were preparing to head out to San Dionisio del Mar—a town at the center of the struggle against the Mareña Renovables wind project—when two state police pickup trucks began circling the school every twenty minutes until the bus left. Driving to San Dionisio, the bus navigated arid landscapes, drove through Unión Hidalgo, passed wind turbines around which despite being the dry season, water was flooding sections of the land. The caravan eventually arrived in San Dionisio where a large group greeted the caravan warmly.

In town, I met another doctoral student who was finishing up their fieldwork. As we discussed their experience living in the village and the conflict between the local PRI and the People's Assembly of San Dionisio del Mar, the issue of narcotics trafficking emerged and the doctoral student explained how they observed an increased violence and drug distribution in neighboring towns since they arrived.

Shortly after this conversation, the caravan headed out to the Laguna where we walked along a narrow sand/rock bar which submerges slightly and is near the Barra de Santa Teresa—the sandbar that divides the Laguna Superior and Inferior. Here people talked together, listened to those giving us a tour of the area, and meanwhile everyone enjoyed the landscape that the resistance had been fighting to defend from wind energy development. San Dionisio del Mar was not the direct access point to the Barra de Santa Teresa (Barra); the town of Álvaro Obregón actually was. The Barra, however, fell within the jurisdiction of the San Dionisio. Miguel López Castellanos, the mayor of San Dionisio affiliated with the PRI, had sold the communal land without consulting the town to Mareña Renovables (Smith 2012), which had recently changed its name to Eólica del Sur. Although Mareña Renovables claims to have paid 20 million pesos (approximately USD 10 million), Lopez only acknowledges 14 million (Smith 2012). This discrepancy caused outrage against the wind energy project, leading people to organize, blockade and file a court injunction (*amparo*) to stop wind park construction. The injunction was enacted on 7 December 2012 by the Seventh Federal Court judge in Salina Cruz (Petersen 2012).

In the evening, the Peoples' Assembly of San Dionisio del Mar arranged a talk during which they stated that their struggle against the wind park had not stopped, even after the court injunction, as the wind company continued trying to enter the Barra. One member of the San Dionisio assembly went on to explain how local politicians had been bought off by the companies and were working to allow the wind company to enter the town. Intense fighting had broken out in the town to prevent the entry of federal authorities who were trying to enforce elections in San Dionisio del Mar. The community, viewing these elections as legitimizing political corruption and the construction of the wind park, refused and boycotted the electoral process (SPIAZ Dec. 2014). A pamphlet written over a year earlier gives an idea of the Peoples' Assembly and their position:

> They came to conquer us over 500 years ago and with tricks they got us to give them gold in exchange for mirrors. In 2004, they returned with another form of conquest [wind turbines], they deceived us to sign a contract to profit from our lands, stripping our brothers from their homelands with the promise of progress, which in its blindness has affected many. Now that we have awakened, they call us a MINORITY, They CALL US REBELS, THEY CALL US DRUNKARDS, they use fake legislation to try to make us kneel with the promise of creating jobs and the chance of earning a crumb of bread, they toy with the hunger of our people, they try to buy consciousness and create conflicts between us (Marquez 2013, 3)

A similar message was echoed at the elementary school in Ixhuatan. After the talk, the nightly caravan performance ensued, and before bed, the caravan grouped together to discuss the complications of this conflict, the political parties' involvement and the position anarchists were going to assume. I began to learn about the variety of factions within the (Indigenous) resistance movement, some of which were less critical of electoral politics and authoritarian leftism (of the Lenin-Maoist variety) than the anarchists. The discussion also included how regional political parties were flip-flopping between negotiating and resisting the wind parks. While the political terrain was complicated, the anarchists preferred working with the local groups and assemblies that were in rejection of not only the wind companies, but also all the political parties.

The next day the caravan continued to Álvaro Obregón, the town located at the entrance of the Barra, which would be the caravan's last stop. On the caravan, I had met people who had told me about this town, about how they had fought multiple battles against the companies, state police and later, the political parties and their armed cadre—'*los Contras*' (see Chapter 4). A year earlier, the people in Álvaro Obregón took over the town hall to form a *cabildo comunitario* or community counsel based on an older form of (religiously) inspired Indigenous governance—*usos y costumbres* (see Gross 2015; Manzo 2011). This literally translates into 'practices and customs',' yet entails a consensus-based system centered on town assemblies and supported by legislation in the state of Oaxaca (Stephen 2005). While I read a lot about San Dionisio del Mar (Smith 2012; SIPAZ 2013; SIPAZBlog 2013–2014; Nahmad et al. 2014; Howe 2014; Bessi, et al. 2014) and other projects in the *Istmo* (Dyer 2009; WDM 2011; Vance 2012; Simon 2013; Juárez-Hernández and León 2014), the community of Álvaro Obregón was known for forming a *policía comunitaria* (Communitarian Police) to stop the construction of the Mareña Renovables wind project—a topic discussed at length in Chapter 4.

Álvaro Obregón, like San Dionisio del Mar, is surrounded by a semi-arid landscape full of bushes, scattered trees and an occasional hill. Driving into town, I thought about how this land would be viewed by the World Bank, Inter-development Bank and the wind companies as 'available',' 'unproductive' and 'underutilized' (see White et al. 2012; Schutter 2011, 543). As the caravan drew closer to town, the vegetation took on a lively green shade and palm and coconut trees appeared. At the town hall, which had been appropriated by the *cabildo comunitaro*, I was pleased to see coconut trees in the main square, anti-wind-energy murals and a large anarchist 'A' painted on the wall of the central gazebo (see figure I.1). Already I felt an attraction to the place. Álvaro Obregón would provide an opportunity to assess local claims as well as monitor what remained of the battle grounds against wind energy where the locals had so far been successful in stopping or, more accurately, displacing the entire project.

We were soon introduced to the people in the *cabildo comunitario* and heard a speech about their struggle against wind turbine development. Afterwards, using the police municipality vehicles they seized during the peaceful takeover of the town hall, they offered to take groups of us out to see the Barra. Excited to see the Barra, I jumped into the back of the expropriated police truck. I sat on the far end of the truck bench with my leg over the tailgate; the truck had about fifteen people crowded into its cab and bed. As we crossed the town and reached the outskirts heading to the Barra, I saw a late 1990s Chevrolet pickup truck parked at a crossroad

Figure I.1. Part of Álvaro Obregón's central gazebo mural. SOURCE: AUTHOR

ahead. In and around the truck was a group of men dressed in black and dark-blue uniforms. Some held guns, including an assault rifle.[4] They stared as we drove by. Feeling uneasy, I asked who the men were. One of the anarchists familiar with the village replied: 'Those are the Contras'. 'Mercenaries?' I asked. The person replied: 'Yes, more or less'. My field notes explain what happened next:

> Driving out to Santa Teresa sand bar in the Communitarian Police truck, we were followed by the Contras. They followed us at a distance, alarming, some more than others, in the truck. I did not feel immediately threatened, but definitely uneasy. Then as we got to the entrance of the Barra [at the salt refinery], we saw another smaller Contra truck drive past us going the other way. People were worried and we turned around and took a right and drove out to the beach. Here we began collecting rocks, sticks and bottles and began preparing for the worst, and discussing what we would do if they followed us [. . . .] After 35–50 minutes, another communitarian police truck arrived because they heard about Contras being out on the road near us with guns. Then a man told us that he had just heard that people from Mareña Renovables had sneaked into town yesterday in a public bus from Tehuantepec and they were going to do survey work on the Barra. The same person who told us this also said that if they [the Communitarian Police] had known sooner that they [the wind company employees] were out here working, then they would have arrested them and burned their vehicles.

Whether wind company employees were out here looking at the Barra or not, I was never able to confirm, but this was the first and only time during my entire stay in the village that I saw the Contras with AR-15 style assault rifles. Moreover, this experience gave me the impression of a strong and determined resistance in Álvaro Obregón that faced an organized and well-armed opponent, but nonetheless was ready to defend the Barra from wind turbines. This disposition was reinforced and explained later in a public talk to the caravan. A member of the *cabildo comunitario* told us:

> All the people that you can see here are the same people that fight against the wind energy projects. We are not saying "'no'" just for the sake of saying "'no'." We know with certainty that this project is not going to bring any kind of benefit to the community as a whole. The so-well-advertised "'clean energy'" will destroy our indigenous community. We are very conscious about everything. We are simple people as you can see, but we are determined. We are here to fight, to resist. We are rebels. The phrase on the wall of the kiosk [behind us] says: "'We will not even take one step backwards, we will be united and we will not stop'"—this is indicative of our determination. We have similar ideas as you [anarchists]. We are against all political parties. When I started to read about Flores Magón,[5] about all the pioneers and all the European anarchists I felt that we are the same, and all the twenty to thirty persons that are here right now have the same ideas as you.[6]

That night, because of the cancellation of a demonstration in Juchitán de Zaragoza, we returned to Oaxaca City. But leaving the town, I knew where to begin my research. Despite the discomfort of living in a low-intensity conflict area, I knew I had to live in Álvaro Obregón/*Gui'Xhi' Ro'* if I was to truly understand the social impact of wind energy development. Not only to see if the wind companies and police would try to invade, but also to understand why people have taken such a militant stance against the wind companies and their political parties. Given my sympathetic position toward the resistance and my curiosity about conflicts over climate change mitigation projects, Álvaro Obregón was ideal. One week later I would return to Álvaro Obregón and begin my research.

DISCUSSING ANTHROPOLOGY: POSITIONALITY, KNOWLEDGE AND MILITANCY

Anthropology is among the most sensitive and reflective disciplines within the sciences. Among these sensitivities is recognizing and investigating the different, often subtle and nuanced types of violence—political, symbolic, structural, infrastructural and epistemic forms—as they emerge, are transposed and coded into environments and cultures (Scheper-Hughes 1992; Bourgois 2001; 2009/1996; Rodgers and O'Neill 2012; Spivak 1988). Anthropologists are trained to listen with sensitivity, participate and analyze the everyday lives of people, their surroundings as well as their personal and political dilemmas that often takes place in areas of contention, conflict and even war. The tendency of anthropologists to live and engage in the multiplicity of often uncomfortable and conflictual areas for long periods of time separates them from most economists, natural and political scientists. That said, the sensitive and engaged nature of anthropological knowledge production, while impressive, often circulated and deployed as a public 'good' (Erickson 2015/1995), also has a dark and, in my experience, neglected underside that is taken for granted by most practitioners and students of the discipline. One could imagine that this underside of anthropology is so deeply ingrained into the functioning of industrial society that it allows some fundamental premises to remain relatively unquestioned within the discipline. This is why anti-anthropology is necessary for anthropologists who want to generate liberatory knowledge.

For Anthropologists against Anthropology

> It was good to write all this down so outside people know what happened, but in the end, it won't do much because the White people at the university know all this anyway and it doesn't change them'
>
> —Anonymous Indigenous woman from Coast Salish territory
> (quoted in Marker 2003)

This anti-anthropology not only questions the foundational 'home-field' dichotomy that tends to deny the connection between where the researcher originates and the area where research is conducted (Gupta and Ferguson 1997), but extends this self-critical gaze to the university system, the practice of anthropology and knowledge production itself (see Harrison 1991; Starzmann 2016). The three are deeply interrelated and the following proceeds by examining each—university, anthropology, and knowledge—in turn.

The university represents great possibilities and potentials for free academic exploration, self-development and critical thought. In its present state, however, the university system has circumscribed and attempts to arrest the breadth of these potentials. The university is subsumed by market principles, experiencing cuts to social spending amidst the prioritization of the 'practical branches of knowledge'—ones that states and businesses consider useful to the perpetuation of government, enterprise and consumer society (Veblen 1965/1918, 57). In the 1918 classic *The Higher Learning in America*, Thorstein Veblen detailed the fusing of market values with universities and how this was destroying any sense of academic freedom. This trend, advanced by William Dugger (1989), has only intensified with new information technologies and the steady neoliberal restructuring of the university system (Giroux 2014; Hyatt et al. 2015). This restructuring, in practice, entails new labour relationships that create precarity for faculty and staff, as well as debt and uncertainty for students (Heatherington and Zerilli 2016), which coincides with budget cuts, audit cultures and the silencing of radical academics (Nocella et al. 2010). Furthermore, publishing pressures combine with job insecurity to create health complications (Bal et al. 2014). This emerges alongside the securitization of universities, transforming spaces of learning into mazes of locked doors, elevators and identification cards where the Vrije Universiteit Amsterdam is an exemplar, but certainly not an exception. The university is becoming increasingly inhospitable to free academic exploration that is critical of state and market structures. If one's work is not directly tied to profit-generating activities, one's employment becomes increasingly insecure. Rather than working towards creating knowledge for a healthier, freer and genuinely socio-ecological sustainable politico-economic systems or explorations into alternative social, political and ecological developments, universities have prioritized research wedded to population control and market expansion, while simultaneously consolidating the trajectory of industrial development necessitating (speculative) finance, weapon development and extreme energy industries. Simply stated, ideas questioning the source of ecological and climate change outside of market 'solutions' and statist control are relegated to the periphery, are marginalized or left in the realm of academic exercise.

The discipline of anthropology is a colonial art designed to explore and make legible social and cultural differences. Tied to the imperatives of the colonial project of information gathering, cataloging and territorial control (Lewis 1973; Harrison 1991; Shalins 1995; Wofle 1999; Lawless 2009), anthropology has not been a neutral description but the creation and justification of particular worldviews of superiority. At the extreme, anthropology helped provide a rationale for genocidal atrocities in both the colonies and the Third Reich (Wolfe 1999; Hinton 2002; Churchill 2012). It employs the language of disinterested or 'objective' scholarship to affirm bureaucratic, expansive and industrial cultures as the most 'developed,' 'superior,' and, in some cases, as the natural trajectory of personal and collective development. A trend that is not passé, but unfortunately centre-stage as the modus operandi of industrial societies have not changed, nor has the role of knowledge production—even if it has diversified. This is tied to the smaller-scale concern of anthropology projecting its own culture through its research emphasis and interest. Michael Marker (2003, 367) aptly writes:

> Researchers have recently discovered that Indigenous communities are not homogenous and have been fascinated with the internal disputes, contradictions, and gossip about families and factionalism that reverberates in all communities, Indigenous or not. This has really become simply another chapter in the history of the colonial gaze at the Indigenous other.

These words are especially pertinent to concerns and debates around ecological indignity. Anthropologists sidelining or condemning ecological stewardship or patterns of past cultures quickly turn into colonial apologists or, to a lesser-degree, indirectly undermine past (e.g., Rambo 1985; Raymond 2007), present or even future desires of Indigenous ecological re-vitalization (Escobar 2008; 2018). This is not to say there is not Indigenous pollution, but for anthropologists to focus solely on it while minimizing the historical coercion and the organizational systems put in place to systematically mine the environment and people for capital accumulation (Rambo 1985; Raymond 2007) is nothing more than the most violent kind of epistemic violence (see Foucault 1989/1961; Spivak 1988; Marker 2003). In other words, anthropology, as it is situated in the university economy and despite its discursively critical elements, works to sustain the imperatives of state consolidation, control and economic expansion that embody structural violence, institutional racism, environmental degradation and low- and high-intensity warfare. Anthropology itself is a knowledge extraction industry (McGranahan et al. 2016; Harrison 1991) demanding critical self-reflection. In short, there is nothing objective about approaching the world from an institutional or statist perspective often implicit and embedded in anthropological positionality. This means recognizing the failure to question and challenge our confining and ecologically destructive environments—university departments or neighborhoods—that are protected by habit, personal comfort, police and bureaucratic procedure. The exceptions of (some) feminist, anarchist, queer and de-colonial anthropology prove the mainstream rule wedded to the direct and indirect knowledge production geared toward advancing state-corporate political control and industrial development.

The structural violence of universities is accentuated by the controversial, yet ever-present collaboration of anthropologists with the military (González 2015, 2010; Price 2014, 2011; Lawless 2009; McFate 2005; Salemink 2003; Solovey 2001; Farmer 1979), police departments (Kania 1983; Simpson 2014) and marketing agencies (Barnes 2009; Baer 2014). Lutz (2009) estimates that twenty-five percent of all scientists and engineers in the United States work on military projects which do not include the indirect and covert accounting and collaborations pulling in anthropologists—a trend I would speculate is only increasing with university privatization. When governments, security forces and business leaders deem anthropological knowledge 'useful' this undoubtedly exemplifies the dark side of 'public anthropology' (Erickson 2015, 398), where public good becomes synonymous with the interests of governments, security forces and corporate profit (Dugger 1989; Bourdieu 1998). This even extends to new 'Market and Management Anthropology' degrees,[7] which harness anthropological knowledge for corporate sales, public relations and managing consumer opinions. Anthropologists are actively being trained to specialize in the engineering of consumer perceptions in order to apply anthropological knowledge to corporate and brand needs. These processes are long institutionalized (McFate 2005), meanwhile some anthropologists have put forward sharp critiques of those in the field who work directly and indirectly with the military and their associated institutes (NCA 2007; 2009; González 2010; Price 2011; 2014), but the former appear as a minority when these processes continue unabated within anthropology departments around the world. When researchers benefit from the current socio-political situation, lacking a critique of state, market processes and exploitation, it only encourages a bias in favor of the state control and profiteering from the destruction of ecosystems. There are serious structural concerns in anthropology in need of immediate redress which because of the structural determinates of funding and university culture seem unlikely to make it pass the cyclical, trendy and ultimately, in terms of social change, dead-end debate.

This leads to the question of knowledge production and its implication especially in relation to systems of governance. In *The Will To Knowledge: The History of Sexuality: 1*, Michel Foucault (1997, 20–5) gives the example of the 19th century book, *My Secret Life*, written by an anonymous author who, in defiance of religious prohibitions, detailed at length his adventurous, hyperactive and 'blasphemous' sexual life. Yet Foucault points out that the author, rather than undermining the regimenting and prohibition of pleasure, helped to advance the apparatus of sexual regulation by performing, albeit in a different way, exactly what the confessional had sought: to highlight and expose the multiplicity of sexual acts and relationships. His account helped make sex increasingly legible to religious and centralized authorities in their efforts to document, classify and manage 'sexual deviance'. This self-production of knowledge, though defying censorship, provided information 'to be not simply condemned or tolerated but managed, inserted into systems of utility, regulated for the greater good of all, made to function according to an optimum' (Foucault 1997, 24). Knowledge produced from the religious and later scientific gaze on the habits and values of people created the possibility of regulating, harnessing and controlling populations, enabling a type of biopolitical framework to manage the political economy of population. This example serves to show the complexity of power around knowledge and how an individual's defiance can easily aid a larger program of governance by producing knowledge and information about the sensitive, personal and often banal aspects of life. This is exactly why anthropologists have become valued assets of colonial armies, the military, police forces and marketing agencies.

The problem of institutional agendas and researcher positionality emerged acutely for me over four years ago in an ethnography workshop. In a room of about forty students, the professor discussed the foundations for ethnographic fieldwork, explained how to approach people in the village, find 'gatekeepers' to gain acceptance in the village, profile people and identify (vulnerable) people willing to talk to outsiders to construct 'informant networks' for research. It was at this moment, looking at the uncritical gazes of my fellow students, remembering past conversations, ambitions, dispositions and, more importantly, our twenty previous years within schooling institutions led me to realize: the lessons taught in this ethnography workshop were not all that different than workshops on counter-intelligence and undercover policing. Implicit in this ethnography workshop and anthropology in general, similar to Foucault's example, is the uncritical idea that all knowledge production that is approved by the appropriate institutional channels is good, a public service, or at the least, harmless. In England and elsewhere, ethical review boards oversee and judge whether or not research conforms to ethical guidelines. This is intimately tied to the university system's dealings with private business and security agencies, which codes a utilitarian value system into university classrooms and infrastructure (see Giroux 2014). In practice, conversations about ethics in the academy rarely extend to challenging the foundations and infrastructures of anthropology, let alone the university system itself. Anthropology functions as a competing business at the university, seeking 'big money' grants and greater enrolment in competition with other departments and schools. Moreover, uncritical anthropologists with the best of intentions can easily provide valuable information to the state and corporate interests, especially if they underestimate the level of embedded political conflict in their research areas or are unaware of what kinds of information can be used by security forces and resource extraction companies. The third chapter reveals a commissioned study that sought to find 'possible solutions to deactivate the social movements that have arisen around' the Bíi Hioxo wind park outside Juchitán de Zaragoza. Research elsewhere (Price 2011; 2014; Brock and Dunlap 2018; Dunlap 2019) has affirmed social scientists as key components in the process of controlling land and pacifying recalcitrant populations resisting resource extractivism or the imperatives of governments.

Furthermore, some anthropologists work to improve police and military operations with disgruntled communities and native populations at home and abroad (McFate 2005; Simpson 2014; Çankaya 2015). These efforts are supposed to address various levels of systemic political violence perpetrated by the military, police and extra-judicial forces, but this research often serves to reinforce, strengthen and intensify the role and power of these institutional actors. Anthropologists or sociologists who work with police and military give insights and feedback that facilitate organizational growth, efficiency or 'optimums' that work to consolidate the power and legitimacy of an expanding war machine and to manufacture a population docile to economic and governmental imperatives. Then again, as in the United States, where systematic police killings of black men has led the NAACP to request that police wear cameras for accountability (NAACP-LDF n.d.), as if there is not enough video and evidence of police brutality and murders circulating with justice pending, if not denied (see The Counted n.d.). Some of these researchers may intend to make the system more just or fair, working to reduce the deaths of civilians and security personnel and develop lines of strategic communication for police. Yet their work strengthens the institutional powers and helps erase the legacy of injustice, violent terror, slavery and genocide that underlies nation-state formation, economic institutions and the police apparatus (Güven 2015; Gelderloos 2013; Williams 2007; Moses 2002), which diversifies their rule instead of rightfully questioning its existence in the first place. This dark side of public anthropology aids the institutions that enforce the disciplinary order of political economy and promotes what Foucault (2007, 339) calls a 'permanent coup d'Etat' on the population to manage and ensure mass submission to politico-economic imperatives of states and their economic arbiters.

The fault lines within anthropology that aid structural and infrastructural violence are located with the social conditioning and positionality of researchers, which responds to the surreptitious and changing notion of 'public good'. The issue of political positionality in relation to issues of extractivism, colonialism, food systems, patriarchy, police and militarism among other pressing issues requires serious consideration and reflective praxis in one's research and daily practice to subvert the political economy of submission necessary for industrial-scale socio-ecological destruction. All anthropologists should radically question the foundations and purpose of their discipline and, consequently, themselves and their motives.

ANTHROPOLOGY IN THE FIELD: MILITANCY

For the reasons mentioned above, the positionality of the researcher is crucial. To carry an anti-anthropological tension throughout the research is to broaden the relational lens of anthropological work and to promote transparency and ethical honesty in a discipline where ethics are all too often embedded in and/or dissolved for the advancement of state and market-based institutions. This tension entails an anarchistic engagement critical of forces that marginalize and subdue lifeways, cultures and relationships between the human and non-human worlds. Living in a conflict area, as Hoffman (2003) notes, often means taking a side with people or groups with whom one shares interest, concern and even affinity. Taking sides can assist the researcher to navigate a region's complexities as well as the psychological and physical stresses of anthropological research in conflict areas (Nordstrom and Robben 1997). I embraced a version of front-line or militant anthropology. Attempting to generate 'practical and embodied understanding[s]' of the situation and processes being researched (Juris 2007, 166), front-line anthropology questions the divisions and hierarchical inequalities between the researcher and research participants to create engagements to learn about each other's experiences and insights. The Global South should not be homogenized, nor should the existing diversity among Indigenous communities be tokenized. Instead, researchers should find their path to developing collaborative relationships and affinities, even if there is always a risk that relationships can change and new, often uncomfortable, insights can unfold and appear.

The researcher and research participants can be each other's accomplices and even friends. It is crucial to recognize the common interests, goals or our mutual yet different forms of confinement in each other's individual and collective struggles, and how research can be complimentary. Embodying the idea that one is free only when all human and non-human beings are free to experiment and create alternatives to the problems they face in their different lives and social (or anti-social) contexts (see Bakunin 2005/1871). There has been much written on the topic of front-line, militant and activist anthropology (Fals-Borda 1987; Taussing 1987; Scheper-Hughes 1992; 1995; Nordstrom and Robben 1997; Bourgois 2009; 2001; Stephen 2002; Kovats-Bernat 2002; Hoffman 2003; Shukaitis and Graeber 2007; Craven and Davis 2013; Juris and Khasnabish 2013), of which the words of Robbens (1996, 343) replying to Scheper-Hughes (1995) remain instructive: it is 'problematic to view "the oppressed" as a homogeneous grouping, and "doing the right thing" more than ever requires a sophisticated and nuanced understanding of the micro-politics of local situations'. This means not only skepticism toward the state, economic and university structures that an anthropologist inhabits, but also an awareness of the dynamics of conflict in the area of research.

While subscribing to a personalized version of militant ethnography, a number of difficulties arose in the field. When picking a side in a conflict, it becomes important to remain self-critical of political sympathies and affinities. Political affinities can lead to trusting people and their analysis, which runs the risk of being selective and self-serving to one's own feelings and subjectivities, which bleeds into how you read the situation. Relying on others who have a similar political disposition can interfere with understanding the *real micro-politics* and political terrain one inhabits as a foreigner. This can create uncomfortable situations.

The simultaneous identity as activist and researcher—participant observation—can prove confusing to people, even if they are repeatedly informed about one's research. When research participants use information sheets to roll joints, how well do they understand what it says? Anthropological work may not make sense, may seem stupid or superfluous in a rural semi-subsistence village immersed in conflict with transnational corporations and between members of the town. Furthermore, signs of wealth can create envy among people in the village and choosing a house may entail choosing between the families in the resistance that have interpersonal tensions among them. While there are no clear answers to the difficulties of research in conflict areas or the self-critical feelings about the binary of researcher and researched (Juris and Khasnabish 2013), what I found important is to stay true to one's own subjective position, locating friends and being increasingly conscious of how one is transformed into a player in the conflict. Conflicts are not 'black' and 'white,' even if they are archetypally or at a macro level—transnational company versus Indigenous village—presented that way. Conflicts quickly become multifaceted due to the acceptance and by extension, the profitability of large-scale development projects which depend on creating communal divisions and factions to support the development project that becomes interlaced with people's individual desires, tensions and stresses. Said differently, colonialism, and now development, has always necessitated interlocutors and collaborators who cannot be forgotten when analyzing such conflicts. Conquest and land control is more complicated on the ground than it is when popularly conceptualized within oppressor/victim dichotomies. Even if they appear 'black' and 'white' through macro frameworks and identity generalizations, the ambitions, desires and poverties of people tend to make unpredictable situations and unlikely alliances.

LITERATURE REVIEW AND THEORETICAL APPROACH

This research on industrial-scale wind energy[8] development in the *Istmo* draws on various and related disciplines: anthropology, geography, critical agrarian studies and political ecology.

Critical agrarian studies offer an important body of research around two indispensable terms: 'land grabbing' and 'green grabbing'. 'Land grabbing,' a term inspired by Karl Marx (2010/1867), designates the appropriation, enclosing and integrating of large tracts of land under public and private control. The notion rose to prominence in popular media around the 2006-08 food crisis and has since generated a number of special issues and articles concerning the complexities of land control and grabbing (Zoomers 2010; Grajales 2011; Li 2011; Peluso and Lund 2011; White et al. 2012; Borras et al. 2012; Edelman et al. 2013; Grajales 2013; Scoons et al. 2013; Wolford et al. 2013; Hall et al. 2015). Critical agrarian studies have been instrumental in revealing the processes of enclosure and land regime changes. Drawing from this literature, this book examines the process of regularizing land titles to shift the land use from an agrarian land regime—based on farming, livestock and cattle—to that of so-called 'wind farming' were agrarian activities are replaced or merged into co-existence with wind turbines—that appear as enormous steel robots with spinning heads occupying the countryside. Emerging from this literature is the idea that land grabbing was taking on a distinctly 'green' character, giving rise to the notion of 'green grabbing', which has historical precedents in conservation initiatives around the world.

The term 'green grabbing,' coined by journalist Gore Vidal (2008) and advanced by Fairhead et al. (2012; see also Corson et al. 2013; Holmes 2014), is a type of land dispossession and displacement associated with the construction and militarization of conservation national parks (Peluso 1993; Brockington and Duffy 2010; Büsher et al. 2012; Ybarra 2012; Lunstrum 2014; Marijnen and Verweijen 2016; Massé and Lunstrum 2016; Duffy 2016; Verweijen and Marijnen 2018/2016; Huff 2017), carbon sequestration and biodiversity offsetting schemes (Lohmann 2006; 2008; 2012; Gerber and Veuthey 2010; Overbeek et al. 2012; Seagle 2012; Sullivan 2013b; Holmes 2014; Bull and Aguilar-Støen 2015; Brock 2018), as well as green energy projects largely associated with biofuels (Borras et al. 2010; White et al. 2012), which includes the under-researched area of photovoltaic solar (see Siamanta 2017) and wind power. There are few studies that directly address wind energy in relation to green grabbing (see Hadjimihalis 2014; Siamanta 2016), but there are a number of existing studies on wind energy development that grapple with related and overlapping issues. For example, Pasqualetti discusses visibility issues with wind turbines (Pasqualetti 2000) and landscape valuation (Pasqualetti 2011), while Munday and colleagues (2011) explore impacts on the local economy, and others have pointed out the inadequacy of 'not in my back yard' (NIMBY) thinking to explain the emerging conflict over wind energy (see also Phadke 2011; Zografos and Martínez-Alier 2009; Wolsink 1994; 2000). Land contract inequalities, community consultation and wind turbine locations are consistent issues with wind energy projects (Phadke 2011; Wolsink 2007; Pasqualetti 2001).

Taking on the political ecology approach of ecological distribution conflicts (discussed further, below), Christos Zografos and Joan Martínez-Alier (2009, 1728) stress that these are not "market imperfections or bureaucratic obstacles," but the result of planning 'that reproduces top-down decision making' insensitive to local concerns that is triggering opposition (see also Avila 2017). Conflicts also emerge between corporate and community wind energy projects (Munday et al. 2011; Oceransky 2011). Inadequate consultation, unequal benefit sharing and ecosystem disruption have led to discontent and conflict in Scotland (Murphy 2013), Wales (Mason and Milbourne 2014; Munday et al. 2011), the United States (Pasqualetti 2000; 2001; 2011; Phadke 2011), Sweden (Lawrence 2014), Spain (Zografos and Martínez-Alier 2009), India (Mate and Ghosh 2009), Greece (Argenti and Knight 2015; Siamanta 2016), Kenya (Danwatch 2016), Brazil (EJOLT 2016), Chile (EJOLT 2016a) and, of course, the *Istmo* region of Oaxaca (Oceransky 2011; Hamister 2012; Smith 2012; Vance 2012; Simon 2013; Howe 2014; Juárez-Hernández and Leon 2014; Howe and Boyer 2015; Howe et al. 2015; Friede 2016; Avila 2017). While the conflict and degree of discontent varies, evidence is emerging that suggests wind energy development is conducting business in the same way as fossil fuel industries: undemocratic procedures, manipulative land deal negotiations, minimization of the social and environmental impacts (see Downey et al. 2010; Bebbington and Bury 2013; Brock and Dunlap 2018). This business takes on clear colonial dimensions when these projects take place on Indigenous land and disregard Indigenous territorial rights and cultural practices (Baker 2013; Franco 2014; Fontana and Grugel 2016). While the studies cited above have contributed to the literature on political ecology and critical agrarian studies, few have taken on the recent insights from land/green grabbing.

In the special issue on Land Grabbing in Latin America, Saturnino Borras, Cristóbal Kay, Sergio Gómez and John Wilkinson (2012) understand the necessity of expanding the definition of land grabbing while distinguishing seven significant characteristics. The first characteristic is 'Land concentration and "foreignisation" as central features of the land grabbing narrative in Latin America'. Foreignisation is associated with privatization and refers to the acquisition of land by foreigners, while land concentration is the consolidation of private property and ownership by both transnational and pre-existing regional actors. Both processes were happening as early as the 1980s, if not for a previous century in Mexico and other parts of Latin America. Borras et al. (2012) continues to assert that international organizations, such as the Food and Agriculture Organization of the United Nations (FAO), narrowly attribute land grabbing to 2007–08 food crisis that has a much longer history. This leads to the second point: 'Land grabbing can be, and is, carried out by domestic capital, often in alliance with the state' (Borras et al. 2012, 407). In other words, foreign and domestic capital are both active, often working together in grabbing land domestically.

The third point is to show how land grabbing, land concentration and foreignisation occur in various economic markets—both agricultural and non-agricultural sectors.[9] The fourth emphasizes how large-scale transnational investments are driven by domestic policy. This has allowed transnational, trans-regional and intra-regional land grabbing by governments and companies of low-to-medium income countries around the world, notably by the BRIC (Brazil, Russia, India and China) countries.

The fifth point emphasizes that land is being controlled in more ways than outright purchase. The market and existing laws are principal mechanisms that manage and facilitate land access, control and acquisition by national and transnational actors (Ribot and Peluso 2003; Peluso and Lund 2011). Land leasing is one technique that Borras and colleagues (2012, 411) call an 'iconic mechanism of land grabbing' that utilizes 'the long-term lease, usually a 99 year lease renewable for a further 99 years'. This has been the principle method used by wind turbine companies in the *Istmo,* leasing land for 20–30 years with three automatically renewing leases. The exact number varies according to the specific project, but the principle remains intact: one does not have to own the land to use it or control it (see Peluso and Lund 2011).

Notable among these techniques are the use of non-profits trusts, NGOs and state regulation that secure land for conservation parks, timber concessions, hydroelectric and infrastructural projects as well as carbon sequestration markets (Holmes 2014). Land grabbing is not restricted to the binaries foreign-domestic or private-public, but tends to have a multiplicity of formal and informal private-public partnerships (see Hildyard 2016) that work across and between sectors to control land and natural resources.

The sixth characteristic of land grabbing emphasizes the role of the state as an 'active promoter of large-scale land deals, sometimes in support of or in alliance with foreign or domestic capital' (Borras et al. 2012, 411). The state structure and its institutions facilitate and embody various modes and diverse actors (Wolford et al. 2013), and the state's quest for investment, economic growth and developmental expansion generally encourages large-scale land investments. When any development projects are met with resistance by local populations, security forces are usually deployed to manage the process of exploitation, displacement or pacification to integrate people into model village schemes (Wilson 2014), eco-tourism (Ojeda 2012), participatory conservation (Benjaminsen and Bryceson 2012), which has contributed to the national and international rural-urban migration of peoples that, following Marx (2010), is creating a reserve army of the unemployed (Wise and Covarrubias 2008).

The final and related point is that not all land grabs result in displacement and conflict. Some local communities and indigenous groups, or sections of them, do embrace and welcome what others call 'land grabbing megaprojects'. Furthermore, land/green grabs are not as simple as a conflict between local communities and transnational corporations or the state. Different land deals meet complex reactions 'from below' that can pit unions against small-holder farmers, set different indigenous groups against each other, and even divide families (Borras et al. 2012; Borras and Franco 2013; Hall et al. 2015). Such divisions were prevalent around wind energy developments in the *Istmo*. In Chapter 3, I advocate a counterinsurgency warfare perspective for examining the social technologies used to divide and conquer communities, building from and further complicating an already chaotic political terrain, to making development projects politically feasible.

Taking into account these seven points from Borras et al. (2012), we find that green grabbing projects displace and alter local and Indigenous lifestyles, habitation and subsistence patterns even though they claim to promote the 'greater good' of mitigating climate change and advancing social and sustainable development (Fairhead et al. 2012; Sullivan 2013a; Holmes 2014; Bull and Aguilar-Støen 2015). Political ecologists have shown how development projects that disregard local concerns and power dynamics can create ecological distribution conflicts (Martínez-Alier 2002) which arise from local power inequalities, unequal distribution of benefits and negative environmental impacts. Green grabbing is an important theoretical tool that represents the proliferation of ecological distribution conflicts arising from new economic valuations of natural resources, such as the wind, solar and tidal wave energy, among others. Governments are actively 'rolling out' new policies agreed upon by various national and international actors and governing boards to sustain the growth of the industrial economy related to ecosystem and climate commodity markets (Dunlap and Fairhead 2014; Springer 2016). Land/green grabbing and ecological distribution conflicts theoretically guide this research into wind energy development in the *Istmo*. Green grabbing represents the proliferation of ecological distribution conflicts arising from new economic valuations of wind resources, illuminating the coercive nature of sustainable development, its consumption of the land, displacement and integration of populations into economic processes, and in the meantime creating continuity between conventional and 'green' extractive activities—from coal mining to wind turbines. This book thus serves as an exploration into the political ecology of wind energy development.

This research was developed around a general theory of social warfare inspired by Michael Foucault (2003), which was later applied to understanding conflicts arising from climate change mitigation measures that relied on market mechanisms to manage landscapes (Dunlap and Fairhead 2014; Dunlap 2014a; 2019). Advancing thinking around green grabbing, Dunlap and Fairhead (2014) argued that climate change discourse and the green economy instigates or reignites old conflicts, and that the green economy continues a war for the acquisition of human and non-human resources by other means. 'Green' rhetoric is used to legitimize and continue such destructive development in contentious areas, including Indigenous territories, while attempting to cloak the environmental and social impact and the potential to generate conflict. While acknowledging that green technology opens market frontiers, it also argues that green economic rhetoric acts as a pacifying device to make politically-feasible economic expansion and land acquisition under the guise of environmental ethics that make encroaching on Indigenous territory fashionable and politically feasible or, at the least, more difficult to resist with the moral weight of 'stopping climate change' or 'saving the world'. This recuperation of environmental discourse responds to widespread militant Indigenous and environmental activism and is used to thwart resistance to green extractivism and serves as a new tool in the counter-insurrectionary toolboxes of governments. The theory articulated in Dunlap and Fairhead (2014) we might recognize as 'the greening of counterinsurgency' that closely aligns with social war theory (Dunlap, 2019). Here, market-based environmentalism propelled by anthropogenic climate change legislation combined with state imperative for economic growth, foreign investment and environmental security collide to create new and reignite old conflicts over new natural resources that need to be managed and pacified in order to protect new investments in 'green' infrastructure.

With this theory as a framework, I analyze the social impact of wind energy development, focusing on the micro-politics that emerge in the villages facing wind energy development. In the middle chapters of the book (2 to 4) I examine three villages in the region and the conflict (or social war) over the implementation, establishment and protection of renewable energy investments. The idea of social war is drawn from Foucault's (2003, 8) genealogy of 'politics is a continuation of war by other means' and 'war hypothesis' that is largely concerned with social pacification and the manufacturing of political legitimacy (Dunlap 2014a). Lecturing on the Normand invasion of Saxony, Foucault (2003, 104) rhetorically asked the lecture hall: '*How do you expect—they say—a few tens of thousands of wretched Normans, lost in the lands of England, to have survived, and to have established and actually maintained a permanent power?*' The question of social war, and of counterinsurgency for that matter, is how to maintain legitimacy and governance in areas subject to the imposition of new and contested forms of governance, social and environmental relations. The attempt to understand the impact of wind energy development in the *Istmo* benefits from applying a social war perspective. This means examining the micropolitics of different groups and individuals in the region as they praise, negotiate and stand in permanent conflict against wind energy development. Following Foucault, this book asks the question: *How do you expect over a thousand wind turbines—operating, planned and placed in the lands of Mexico—to have survived, and to have established and actually maintained permanent power generation in the coastal Istmo?* While this book examines the social impacts of wind energy in the *Istmo*, how these wind projects arrived, how they were able to establish themselves while simultaneously negotiating and managing a recalcitrant population will be the central focus of this book.

METHODOLOGY AND RESEARCH APPROACH

After about two weeks in Álvaro Obregón, I discovered that the nearest operating wind park was in Playa Vicente. The uprisings and inter-communal conflicts in Álvaro Obregón instigated by wind companies would be relevant to my study of the impact of wind energy, but I needed to understand what it would be like to live close to wind turbines. I continued my participant observation in Álvaro Obregón and began attending the new wind park consultation (*consulta*) in Juchitán. The consultation began in early November 2014 and was an attempt to meet Convention 169 of the UN International Labour Organization (ILO), ratified in Article 17 of the Mexico Constitution in 1990 (Stephen 2005) and required the Free, Prior and Informed Consent (FPIC) of Indigenous populations for the continued construction of wind parks. During the question and answer period of the 5 February 2015 consultation, a local political candidate in La Ventosa spoke to James Anaya, the ex-UN Rapporteur on Indigenous affairs, about the distance of wind turbines to houses:

> I told them about my brother Cesar. He lives in a neighborhood named Altamirano, and a wind turbine is 40 metres from his house and the only thing that separates them is a canal. I have to tell you that my brother didn't invade the wind turbine area; the wind turbines invaded the neighbourhood. He moved to the neighbourhood and 4 years later the wind turbines arrived. It's only 40 metres away. Last time [at the consultation] I invited them [technical committee] to spend one night in that house, to hear the noise. He bought another place near the baseball field and now, 300 metres from his house, wind turbines again arrived. Why are you lying to us? I am giving you evidence and I'm inviting you to go to assure yourself. So, I want a date. When can you come to La Ventosa with me? Anybody who wants can come with me to confirm for themselves that a wind turbine is 40 metres from his house.[10]

Even if the wind turbine companies and ex-UN Rapporteur would not come, I did. One week later this local politician would take me, along with two other doctoral students and a journalist, on a field trip to La Ventosa to see wind turbine distances from houses, talk to its inhabitants, and see the remains of a wind turbine that supposedly exploded due to the high-winds. It was after this day trip and talking to people that I organized my research in three phases of wind energy development.

While based in Álvaro Obregón, I would also conduct research in two other locations within the *Istmo*. The farthest site was La Ventosa, a two-hour bus ride from Álvaro Obregón or forty-five minutes by taxi or car. La Ventosa had been the second town (after La Venta) to become surrounded by wind turbines, and an ideal place to learn about living near these machines (see Chapter 2). Wind energy projects had been operating around this town at various levels for eight years. Between the two extremes of La Ventosa and Álvaro Obregón was the Bíi Hioxo wind park on the southwest outskirts of Juchitán de Zaragoza near the Seventh Section (*séptima sección*) neighborhood. Here, despite widespread social upheavals and barricades to prevent the completion of the wind park on communal land, the project was forced through with the help of police and gunmen, eventually being the first wind park built on the Laguna Superior in October 2014 (see Chapter 3). The Bíi Hioxo wind park is the only wind park built so far, on the Laguna. Researching these three towns within Juchitán County would create room to locate divergent details and commonalities between the wind parks and the affected populations. This approach enabled the examination of wind energy projects in three different phases of development: a long established wind park, a newly established park and a contested site. These sites represent symbols of: (1) wind turbine colonization (Chapter 2); (2) political repression for a wind park development (Chapter 3); and (3) insurrection against a wind project (Chapter 4). This assessment of three different locations experiencing wind energy development in the region permitted access to a wider range of experiences, anxieties and discontents.

My research relied primarily on participant observation and informal and recorded semi-structured interviews. This also included oral history interviews with select individuals in order to learn more about regional history, and cultural and religious practices in the area. These interviews were complemented by secondary materials—newspapers, reports, journal articles and so on—as well as by living and working with resistance groups in the areas affected by wind energy projects. These interviewers were also conducted with an interpreter and a translator. The interpreter, 'Mr. X', had been visiting and working with the resistance in the *Istmo* for just less than two years and was among a series of people to encourage me to live in Álvaro Obregón/*Gui'Xhi' Ro'*. During this process, we became close friends and had to endure a number of uncomfortable situations related not only to the intense interviewing schedule I created, but also the conflictive conditions that surrounded wind energy development.[11]

The following information is based on 114 different recorded encounters and/or interviews in La Ventosa (63), Juchitán (17) and Álvaro Obregón (34). Interviews were conducted between 15 January and 20 May 2015. The number of men and women interviewed were roughly even, while the number of third genders—*muxe*—was substantially lower with only a small handful.[12] The overall town interview imbalance is a result of having to leave the area unexpectedly because I was targeted for repression. This repression was related to my embedding in and participating with the groups in resistance. This included a number of activities such as joining occasional fishing trips in different sites; attending a series of *Vela* ceremonies, pilgrimages and public events; and in the case of Álvaro Obregón/*Gui'Xhi' Ro'* embedding with the *policía comunitaria*. Other interviews such as oral histories and follow-up interviews bring the total interviews to 123. My research was also based on recorded and translated FPIC consultation transcripts and other recorded public events concerning wind energy projects as well as my notes to conversations and interviews with people who did not want to be recorded because of the sensitive discussion content. Anonymity is particularly important for this research, in which research participants are referred to by gender or by means of broad identity descriptors—for example 'farmer,' 'fisherman'—or aliases related to animals to which they had existing relationships or affinity.

BOOK STRUCTURE

The goal of this book is to understand what industrial wind turbine development really implies 'on the ground'. Renewable energy is among the last hopes of the industrial economy (Lohmann 2009), and many consider it a positive step away from fossil fuels, coal and hydraulic fracturing. Meanwhile the Inter-governmental Panel on Climate Change (IPCC) hopes that more than twenty percent of the world's electricity demand will be met by wind energy by 2050, while the US had, pre-Trump, hoped to reach this goal in 2030 (Tabassum-Abbasi et al. 2014). These factors mean that wind energy is likely to increase across the world, yet little is known—at least in the mainstream public—what this practice actually implies. This research was designed to study wind energy development projects in three distinct phases of one project in operation for over eight years, one project running for one year by the Lagoon and another not yet in operation because of militant opposition to stop the imposition of wind turbines and the life it would bring.

Chapter 1 provides context by giving a brief history of the Isthmus of Tehuantepec region (*Istmo*), Oaxaca Mexico. I offer a brief history of the *Istmo* from pre-colonial time up to the arrival of the wind energy projects. This chapter highlights important events from uprisings to land resolutions, patterns of colonial collaboration, and negotiation. This includes discussing important regional figures like Che Gomez and General Heliodoro Castro Charis, development projects such as the Benito Juárez Dam as well as the emergence of The Isthmus Coalition of Workers, Peasants and Students (COCEI). The intention of this chapter is to provide a brief historical background about the political fault lines and dispositions within the *Istmo*, where now wind energy is the latest chapter in this region's history of negotiating foreign development.

Chapter 2 discusses the experience of wind energy development in La Ventosa, and it underscores how people in the area experience wind energy development. People were concerned about the little or no social or collective benefits from the wind turbine projects and how it was only the land owners, political authorities and their associated networks that profited from the projects. Instead of social benefits, the wind energy projects brought discomfort to those living close to wind turbines, social divisions and a type of rural gentrification in which the arrival of wind energy bureaucrats and workers influenced local lifestyles and habits and triggered a rise in land, rent, food and electricity prices, increasing inequality and income gaps between residents. Additionally, there were also significant health complaints of nausea, hypertension, insomnia and chronic headaches reported that resonated with other accounts of the controversial wind turbine syndrome.

Chapter 3 documents the takeover of and resistance to the construction of the Bíi Hioxo wind park on the outskirts of the seventh section neighborhood of Juchitán, detailing the repressive techniques employed by state, private and informal authorities against popular opposition to the wind park's construction on communal land. The use of 'hard' and 'soft' counterinsurgency techniques to marginalize and pacify resistance to enable the completion of the wind park will be closely explored. The environmental concerns raised by locals living and working around the wind park are also discussed, before concluding that the Bíi Hioxo wind park has caused violent social divisions, damaged subsistence and cultural practices with severe consequences, albeit subtle and slow, implications for Indigenous populations in the region.

Chapter 4 deals with the struggle to defend the *Barra de Santa Teresa* from wind energy development. This is a story of resistance and its complications, providing a glimpse into the reality of wind energy development, but also the complicated micro-politics of land acquisition, conflict and unrest faced by a semi-subsistent community. This section begins with a chronology and the community's reasons for rising up against the wind companies and the political parties with the peaceful takeover of the town hall. Divisions in the community and the complicated relationship between protecting the sea, fighting for autonomy and dealing with attacks from a rival faction working with the political parties and the wind companies—the Constitutionalists or *Contras* will be discussed. The chapter also relates observations and accounts of the wind turbines' environmental impact. The research in the area ends abruptly with my decision to leave the community as I was being targeted by a hired gun and the town was preparing for the possibility of a military siege to force elections on the town.

Chapter 5 discusses the recent attempt by the Mexican state to follow constitutional guidelines to provide the people of the *Istmo* with Free, Prior and Informed Consent (FPIC) to further wind energy development. In November 2014, the first FPIC consultation was called for the Eólica del Sur wind project in Juchitán. Lasting eight months, the consultation was responding not only to the UN International Labour Organization's (ILO) convention 169 that Mexico signed in 1990, but also addressed the widespread uprisings against wind energy projects in the region. This chapter begins with a review of FPIC literature, followed by sections examining the consultation in Juchitán, its spatial layout, emerging discursive positions and the repressive atmosphere. Then I analyze the discursive techniques deployed by the FPIC technical committee (TC) which—despite unanswered questions and popular opposition to the wind energy project—granted project approval on 30 June 2015. The chapter concludes that the FPIC consultation undermined Indigenous autonomy, reinforcing a context of substantial political and economic asymmetry between state, corporate and elite interests and Indigenous fishermen and farmers. The consultation reinforced state power while serving as a marketing platform for development projects, and created an illusion of real dialogue, negotiation and democratic decision making.

In the sixth and final chapter, I argue that wind energy in the *Istmo* furthers industrial processes of grabbing Indigenous land and exposing people to the subsequent social and environmental impacts that threaten locals' livelihoods and cultural values and risking the continuation of colonial genocide via wind energy. A definition of colonialism will be outlined in order to identify the continuation of the colonial project and to understand its relationship with wind energy development. The literature on colonial genocide studies, discussing the relationship between the colony model, bio-disciplinary power, accumulation by dispossession and the intention of market operations will be reviewed as well. It will confirm that green market processes in the *Istmo* are rapidly expanding the colonial model and are part of a 'slow industrial genocide' of cultural and biological diversity in the name of mitigating anthropogenic climate change (Huseman and Short 2012). The chapter concludes that the environmental destruction entailed in wind energy development threatens to go unnoticed in the shadow of extreme energy extraction and production associated with oil, hydraulic fracturing, coal and nuclear power.

The book's conclusion summarizes the findings and reflects on the rebranding and intensification of industrial capitalism with renewable energy technologies in general and wind energy in particular. It briefly examines shifts in the technological organization of corporations between the two World Wars, describing how these changes in corporate organization allowed the possibility of infinite organizational growth and profiteering, a shift that retains heuristic meaning when examining the diversification of energy production between traditional fossil fuels and renewable energy systems. This process of technological change is highly dependent on advances in recycling rare earth minerals and other metals along with new changes in promoting industrial sustainability, which embodies the tension and creates the possibility for renewing capitalism and spreading industrial infrastructure to every corner of the earth. It also frames a discussion about resistance from environmentalists who narrowly focus on carbon and limit their criticisms to the fossil fuel economy. The distinction between fossil fuels and renewable energy is misleading in that wind energy requires intensive processes of extraction and is dependent on fossil fuel technologies, which is why it is more accurate to describe renewable energy as Fossil Fuel+. Concerned citizens and environmental activists limit their critique of industrialism to fossil fuels, manufacture a politics that leaves the structural oppressions embedded in this system untouched and have the possibility of strengthening the industrial system through demands for wind energy and other renewable technologies. When industrial capitalism and its ongoing assimilation processes are met by acceptance or by reactive politics that attempt to problem-solve the fossil fuel economy by uncritically and unapologetically promoting and endorsing renewable energies, such 'critiques' risk intensifying the regimes of production, consumption and state control, thus renewing the destruction inherent in the industrial-capitalist system.

NOTES

1. Elementos para la Promoción de la Energía Eólica en México.
2. Most people I talked to knew that the government, business, police/military and the Narcos were "'lo mismo'" (the same). For more references, see Dawn Paley's (2014; 2015) *Drug War Capitalism* and Guadalupe Correa-Cabrera's (2017) *Los Zetas Inc*.
3. The mime act was exceptionally funny and was a great way to break down barriers and play within the community. I would like to recognize how this act brought joy and laughter to the communities visited while also playing with conservative social norms. This performance was magnificent and a pleasure.
4. After discussing with people, they had at least one AR-15 or Galil assault rifle, at least one shotgun and possibly another type of rifle like an M-1 Carbine.
5. For English see C. Bufe and M.C. Verter, eds. *Dreams of Freedom: A Ricardo Flores Magón Reader* (Oakland, CA: AK Press, 2005).
6. Public Talk, 17 January 2015.
7. http://www.sdu.dk/en/uddannelse/bachelor/market+and+management+antropology.

8. The label to describe wind projects has been criticized by people within the wind industry as 'emotionally laden' and 'propaganda terminology', which is not the intention of using the term here (Jeffery et al. 2013). Industrial wind turbines (IWTs) are intended to reference the scale and construction of an individual turbine—it is a product of industrial processing, manufacturing and construction. This is opposed to a small scale or even a 'do it yourself' (DIY) wind energy system appropriate for homes, small-farms and communes (Piggott 2009), which tends to have a different, human-scale relationship with the user. Remembering their industrial-scale reminds people how wind turbines will change landscapes and ultimately how people will relate to them and what locations would be appropriate for their construction. If this 'industrial' aspect is masked or hidden in any way, then discursively, it promotes a hidden agenda that has the propensity to result in social conflicts. Claiming that the word industrial wind turbines (IWTs) is 'propaganda terminology' is attempting to fortify a discursive deception, indicating the future possibility of disagreements and feelings of betrayal when wind turbines are built next to ranches and homes. And it was these problems, among many others that sparked widespread resistance against (IWTs), internal conflicts within towns and families that this book investigates and discusses.

9. Corn, for example, is grown for food as well as biofuel.

10. Q&A FPIC Consultation, 5 February 2015.

11. Thanks again to Paul for making the time to work with me on translating these interviews.

12. This might be called transgender, but more accurately is a third gender that takes on a variety of important social roles in the community (see more description Rubin 1997, 232–33). Considering the patriarchal and sexist culture, there is a surprising level of tolerance for muxe's, but discrimination is still present.

1
Welcome to the Istmo: A Brief History of Politics, Conflict and Development

Before discussing my experience and research findings about the social impact of industrial-scale wind energy development in the *Istmo*, it is important to provide some context. The following is a historical genealogy of the *Istmo*, which highlights important events, conflicts and characteristics of the region from pre-colonial times to the arrival of the wind turbines.

The Isthmus of Tehuantepec (*Istmo*) region is located in the southwest state of Oaxaca, Mexico. The Isthmus is one of the narrowest strips of land separating the Pacific Ocean from the Gulf of Mexico and was a candidate for a trans-oceanic canal before the completion of the Panama Canal in August 1914. The *Istmo's* west Pacific Ocean is home to the Laguna Superior and Inferior which lie near the waters of the Gulf of Tehuantepec. Only 137 miles (220 km) of land separate the Gulf of Tehuantepec from the Gulf of Mexico (EB 2015), a geographical feature that contributes to the *Istmo's* unique character, especially its powerful wind. The close proximity between these two bodies of water, the Sierra Atravesada Mountains and the Laguna all create a strong and constant wind that has attracted wind energy companies to the area for the last twenty years.

FROM CONQUEST TO WIND TURBINES: COLONIAL INCURSION, ADAPTATION AND RESISTANCE

The *Istmo* has a rich and conflictual history. The state of Oaxaca is the most culturally and biological diverse state in Mexico (Robson 2007) and is home to sixteen different indigenous languages (Stephen 2013). The *Istmo* in particular is home to five principal indigenous groups: the Zapotecs, Ikoots (Huave), Zoques, Mixes and Chontals. The Zapotecs remain the dominant ethnic group in the region. Scholars

SOURCE: CARL SACK.

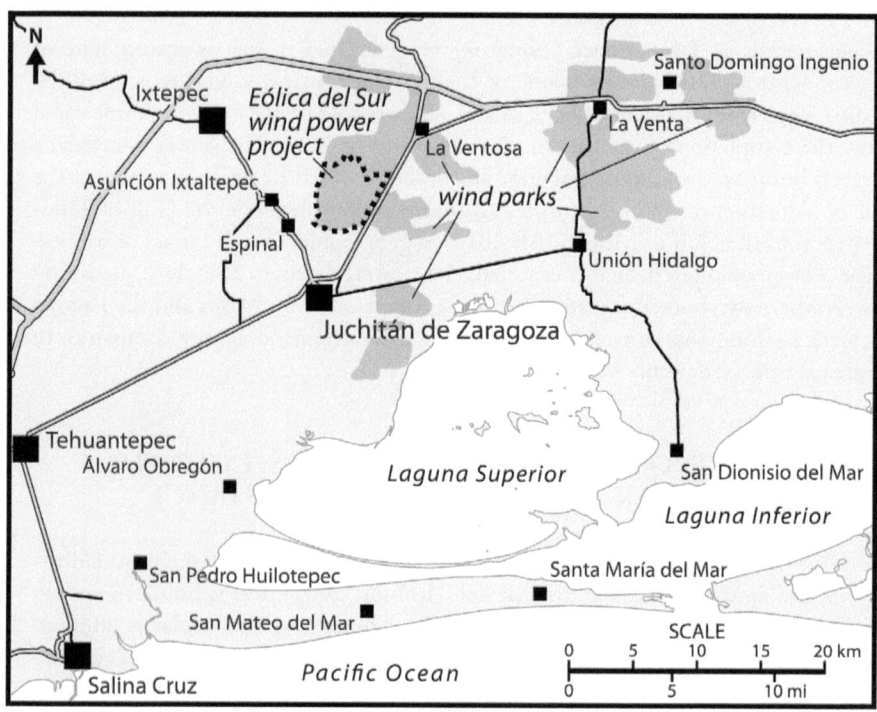

SOURCE: CARL SACK.

are not certain how the Zapotecs came to inhabit the area, but some have theorized that they migrated in the mid-fourteenth century from the highland valley where Oaxaca City sits today (Campbell et al. 1993). Monte Albán is a UN heritage site that rests in the northern hills of Oaxaca City and thrived as a Zapotec pre-colonial Mesoamerican city from 200 to 900 A.D. According to John Tutino (1993, 42–3), it was 'ruled by an elite of warlords and priests and sustained by the surrounding peasantry', justifying their coercive rule as the natural path set forth by the 'deified forces of nature'. The imperial Zapotec civilization eventually spread to the present day *Istmo*, where they absorbed local indigenous groups into their class, caste, gender and age system, displacing them from valuable lands and natural resources (Campbell et al. 1993). Although Zapotec society aggressively seized land and extracted tribute, Judith Zeitlin (1989) notes that the Zapotecs, Ikoots and others in the region coexisted for over a century before the arrival of the Spaniards and after their conquest of the Aztecs in 1521 (Campbell et al. 1993).

Arriving in the *Istmo* in 1522, the Spanish found a deeply stratified class society and a man named Cosijopi—'the last Zapotec lord of Tehuantepec'—who had a diplomatic disposition, and negotiated a new alliance with the Spaniards (Tutino 1993, 45). Cosijopi was soon baptized and took on the name of Don Juan Cortés. He became 'the first colonial broker between European rulers and the peoples of Tehuantepec' (Tutino 1993, 45). Isabel Altamirano-Jiménez (2014, 181) notes that as 'the Zapotec chiefdom was weakened, Indigenous communities used colonial courts to reclaim territorial rights that, in some cases, had never existed'. The Zapotecs tried to adapt to colonial rule, becoming 'colonial intermediaries' and began a process of 'selective borrowing', which demonstrated the culture's capability to respond creatively to the external imposition of colonial rule (Tutino 1993, 45; Zeitlin 1989, 26). Zeitlin (2005) and Altamirano-Jiménez (2014, 181) argue that colonial influence 'reinforced more horizontal relations among community members' struggling with the breakdown of the Zapotec civilization. This was accomplished through a juridico-political system centred on land-owning heads of households known as *cabildo* established across 'New Spain' by the Spaniards. The cabildo was the native municipal government that worked in accordance with repartimiento (the Spanish system of forced labor and production), the Catholic Church and the legal system (Yannakakis 2008), all of which formed to legitimatize the structure of conquest. 'The average degree of intervention to which cabildos were subjected did not destroy independence of action', writes historian Fredrick Pike (1958, 157–58), the cabildo 'had considerable ambit in which to experiment with certain of the forms of self-government'. The *cabildo* eroded local power holders and 'gradually replaced the indigenous *caciques*' (Zeitlin 1989, 52). This was reminiscent of an important technique of colonization developed in the Roman Republic known as 'equalitarizaton', that sought to 'convince inferiors that a little more equality for them would do them more good than much greater freedom for all' (Foucault 2003, 145).

Zapotec Christianity emerged as a cultural adaptation in response to Catholic missionary work, which taught local Indigenous groups to internalize their oppression

(Zeitlin 2005). The Zapotecs appropriated Christianity and combined it with their land-based belief system in order to continue their relationship with the natural environment so intrinsic to their health, agriculture, and fertility. John Tutino (1993, 46) contends that 'Zapotec Christianity became dominant in the annual cycle of community rituals that integrated life in Tehuantepec communities in the colonial area and beyond'. The 'indigenization of both Catholicism and governance practices resulted in a resignification of the so-called "cargo system" of communal labor that considered civil and religious work as communal labor' (Altamirano-Jiménez 2014, 181). Since this time, Zapotec cosmology has adapted and survived within the shell and the cracks of the Christian Church (see Gross 2015), which has created a multiplicity of interpretations and nuances that still survive and have been transformed into tools of resistance against the arrival of industrial wind turbines today.

The Isthmus served in colonial times as a strategic position for maritime commerce. An initial interest of the Spaniards and their indigenous intermediaries was to use the Laguna for ship-building and maintenance with native labor. The *Istmo* is also home to gold, lumber, pitch and salt (Zeitlin 1989). Zeitlin (1989, 40–1) argues that the three main forces of change in the late 16th century were (1) the competition for important natural resources; (2) the removal of natives from community-based activities and labor; and (3) the acculturation of people into colonial behavioral patterns. An important factor in these changes was the shift to cattle herding that spawned a ranching boom, where just two estates in La Ventosa and Las Cruces held more than 12,000 heads of cattle (Zeitlin 1989). *Istmo* Zapotecs consolidated power within the colonial culture, with elites joining the colonial commercial economy, others maintaining a solid subsistence base and still more remaining at the margins of the colonial economy, strategically negotiating their engagement (Tutino 1993). While colonial control steadily declined, African slaves working on estates were freed and given the status of 'free mulattoes', who would integrate into regional Zapotec culture (Tutino 1993). Towards the end of the 17th century, the Spanish decided to double the taxes on the people of the *Istmo*, who responded with outrage. The Spanish Crown tried to punish them, but they rioted, initiating the Tehuantepec Rebellion of 1660. The Zapotecs could not be pacified. Don Juan de Avellán, the man behind the tax, was killed and the region was in an insurrectionary state for about a year, while the Spaniards sent emissaries to try to recruit allies and collaborators in the region (Tutino 1993). Magdalena María, a local woman, was arrested after the uprising and 'convicted of having sat on the dead Spanish *alcalde mayor* and pounded his corpse with a stone while scornfully reproaching him' (Rubin 1997, 29). María would soon contribute to the legendary accounts by outsiders that Istmeños women were fierce, rebellious and cunning. Over time, as the rebellion began to subside, Zapotec power diminished in the Central Valley (near Mitla and Zachila) and Tehuantepec alongside neighboring Juchitán de Zaragoza became the hub of Zapotec culture and power in Oaxaca (Tutino 1993). Meanwhile a silver boom struck, and indigo became a primary export—it was made with the cochineal, an insect that is crushed to produce a high quality red die (Binford 1985). The resultant rise in commercial import and export coincided with population growth that later dropped off with the great famine of 1785–1786 (Tutino 1993). 'After three centuries of colonial rule', writes Tutino (1993, 53), 'catastrophic depopulation followed by a slow demographic recovery, Isthmus Zapotecs maintained a cultural tradition built upon a history of creative adaptation to colonialism and resistance to what they defined as colonial excesses'. This is a history we can see repeating itself in the *Istmo* today.

The Mexican State: Insurrection and Caciquismo

The history of insurrection and uprising only increased with the slow formation of Mexico. Gaining its independence in late September 1821, Mexico suffered near economic collapse from the centuries of Spanish exploitation and resource extraction, but the people persisted when adapting to the new political formations of Oaxaca State (Tutino 1993). The colonial power vacuum was opened; people seized positions of power and attempted the ambitious project of unifying the newly forming Mexican nation-state (Smith 2009). Underneath the tyrannical tendencies of nobles, governors and presidents, a liberal vision was brewing that saw a society ruled by private property, commercial production and a separation between church and state. On the other hand, especially in the *Istmo*, there was the timeless desire for independence, and regional and individual autonomy. The latter position resented any state impositions on communal life, which was a stance shared from local barons to subsistence farmers, who recognized state intervention as a common enemy even if there were internal disagreements among regional actors. The *Istmo* tended to express the latter disposition: locals regarded outsiders—'Oaxacan [City] Mexican European, and US—as perpetually seeking to impose economic and geopolitical claims on the Isthmus', while outsiders viewed the *Istmo* as 'foreign, lawless, and violent, attributing to women in Juchitán both arrogance and seductiveness' (Rubin 1997, 36). In 1829, faced with conscription by the Mexican Army, the people of Juchitán—*Juchitecos*—refused to be tools of the Mexican state, and the women rallied the entire town to occupy government buildings and to free those forced to serve in the military (Rubin 1997). In 1834, Juchitecos would take up arms behind Jose Gregorio Melendez to reject the property rights of foreigners, which would later manifest in people refusing to recognize salt flat concessions sold by the Oaxacan government in the 1840s, where locals maintained a combative attitude toward the owners (Tutino 1993; Rubin 1997). Under Melendez, people would continue to pasture their cattle on the lands of barons and government officials, while raiding salt flats and systematically violating property and tax laws. Zapotec resistance would be diverse, taking the form of 'petitions, land invasion, illegal use of private salt flats and pasture, smuggling, and attacks on persons and businesses, in addition to armed rebellion' (Rubin 1997, 32–3). Their stubbornness won them success, eventually leading President Benito Juárez to try to integrate the rebels' strength into the Mexican government by appointing Melendez to the head of military forces in the region. Melendez accepted this power, not long after declaring the *Istmo* independent from Oaxaca State.

When the Mexican Revolution was on the horizon, the Istmeños were on the frontlines. Porfirio Diaz's (1876–1911) autocratic rule, justified by the quest for political and economic stability, became a central symbolic grievance among competing factions—liberals, anarchists, leftists and a diverse range of communities seeking autonomy and self-determination—which ignited the Mexican revolution (Gibler 2009). Diaz's policies and his oligarchy's 'politics of land grabs,' that by 1906 seized 49 million hectors, mixed with his systematic corruption and heavy handed repression, made revolution inevitable (Gibler 2009, 36; Smith 2009). Attempts by the Mexican government and land owners to privatize communal lands and salt flats spawned Istmeño discontent and accelerated the rhythm of insurrection against Porfirio. Historian Letica Reina argues that the revolution started early in the *Istmo* with the armed revolts of 1872, 1876, 1879 and 1880–1882 (Smith 2009, 28). A notable historical figure leading these rebellions against the imposition from Oaxaca City was Jose F. Gomez, who worked to establish strategic diplomatic negotiations to further *Istmeño* independence, until he was assassinated in 1911 (Campbell et al. 1993; Smith 2009). The *Istmo*, specifically the areas around Juchitán, had the second largest armed movement after Zapata and his followers. The rebel *Istmeño* fought to assert their 'fierce commitment to self-government and local autonomy' (Rubin 1997, 28), which continues in various degrees today against the imposition of wind energy development.

The revolution initiated land reforms, where the legal concepts of *ejido* and communal land appeared in Article 27 of the Mexican Constitution. The *ejido* system transformed abandoned or foreign-owned land into lands held under communal tenure by organized associations that also established *comunidades agraris* (agricultural communities). This was done with the aim to satisfy the demand of landless peasants, who watched and experienced the *haciendas*—feudal agricultural estates—usurp their land, natural resources and labor. In the years 1917–1934, after the revolution, the Agrarian Department recorded 112 *ejidal* decisions that allocated 128,828 hectares of land. Under President Lázaro Cárdenas (1934–40) this number increased to 262 land allocations of around 374,735 hectares[1] (Smith 2009). The *ejido* provided land for farmers to cultivate, but not to buy and sell. This changed after alterations to Article 27 in 1992 that made possible the privatization of *ejidos* and established the PROCEDE[2] land certification program (Assies 2008). Communal land is related to pre-colonial land claims and governed by the community assemblies—*comuneros*—who are traditionally all male and govern these lands based on varying rules and relationships according to regional customs and practices (Stephen 2002).

Communal land recognition was written into the 1917 constitution, but the notion of *comunidades agrarias* was not formally recognized by the federal government until changes to agrarian law in the 1940s and later in 1966, receiving special status as 'private land from communal origin' (CODIGODH 2014). In this respect, it is important to note that while 'only about 5 % of the land in Mexico was held under communal tenure in 1960', '38% of the Oaxaca land was so administered' (Binford 1985, 180–81). Oaxaca has the largest Indigenous population in Mexico as well as the highest concentration of communal land, which is regarded by the Mexican state as a barrier to economic development (as President Peña Nieto's words in the Prologue demonstrated). This has resulted in a variety of legislation, programs and interventions to better manage and integrate these lands and people into the Mexican political economy (Stephen 2002; Payan and Correa-Cabrera 2014; Bryan and Wood 2015). All of these different forms of collective property systems—*ejidos, comunidades agraris or comunales* (communal lands)—were regarded as social property (Stephan 2007, 124–25), but an important distinction between *ejidos* and communal land is that *ejidos* under Article 27 granted 'direct ownership of all natural resources' below *ejido* topsoil to the Mexican state (MG 2007, 19). This combination of state resource control and agrarian land concessions inherent in the *ejido* can be read as working to pacify revolutionary tensions in order to promote state unification and export-oriented agriculture (Tutino 2007), which would continue to be utilized over time as a tool by various political parties, notable among them the COCEI, to mobilize and strengthen their power base (see Campbell 1994, 276).

New power vacuums formed after the revolution. The revolution gave birth to the National Revolutionary Party, later renamed the Institutional Revolutionary Party (PRI), which would dominate Mexican politics for the next seventy years and was devoted to state unification. The Mexican state worked to pull together and govern a diverse array of states, regions and people, who struggled to maintain their autonomous or semi-autonomous character while simultaneously negotiating the imposition of the Mexican government, modernity and capitalistic economic relations. What emerged in Oaxaca after the revolution was a state system constantly battling with *caciquismo*, a style of politics centred around dominant political bosses or regional power brokers who negotiated or battled with external influences. These *caciques* historically maintain power through clientele relationships, often relying on violence and electoral fraud to maintain their regional position of power in a crude, naked form of politics. Internal political power struggles were rampant after the revolution, and Genaro V. Vásquez managed to fight his way into governorship of Oaxaca, where he moved away from using brute force and instead 'attempted to divide opponents with assurances of regional independence and then draw them into his governing camarilla'—a political network who shared a common interest (Smith 2009, 39). This move formalized *caciquismo* by capturing the 'hearts' and 'minds' of local *caciques* that had been entangled in post-revolutionary violence. He also started to articulate and propagate the idea of *Oaxaquenismo*, a nationalist construct that was designed to try to unify the people to collectively identify as Oaxacans (Smith 2009). In an interview, a local 'organic historian' drew continuity between Spanish rule, the Mexican state and more recent techniques to integrate different Indigenous groups into the state project in Oaxaca:

> During the colonial years the Spanish had a lot of trouble getting all of the communities to go into their markets; it was a ferocious fight because they kept trading amongst themselves in non-monetary ways. So the Spanish tried to monetize the economy for three hundred years and they couldn't. Then the Mexican Liberal state could not get

Indigenous people to embrace private property and are still trying. After the war or what is called the Revolution, the Mexican state literally collapsed and there was no state for a while during the 1920s. One of the big tools they used to integrate communities was the creation of *municipios* [townships[3]], but in Oaxaca they had different requirements to become a township than the rest of the country. In Oaxaca the requirements were considerably less, here to become a township all you needed to have was a population of 500. So with five hundred people you could have your own township and if you had less than five you could sign up as an *agencia municipal* [municipal agency], which they have here [in Álvaro Obregón].

The allure of legalization and official recognition by the Mexican state is rife with negotiation, conflict and communal divides and still continues today with ideas of remunicipalisation of districts (*remunicipalización*, see Solano and Mayor 2007).

In the *Istmo*, an important figure and popular *cacique* was General Heliodoro Castro Charis, a Zapotec who was politically active in his teen years against the minority 'Reds' who collaborated with Oaxacan elites, controlled Juchitán and were characteristically *mestizo* and regarded as outsiders (Rubin 1997). Charis would go on to fight in the Mexican Revolution, eventually leading a battalion of Juchitecos in support of the national leader Álvaro Obregón, and as Obregón rose in rank, so did Charis. In the 1920s and 30s support for Charis grew. As a 'Green' he fought for local autonomy, well-being and continued the tradition of resistance to outside intervention from Oaxaca, Mexico City and beyond (Rubin 1997). Charis became the powerful regional *cacique* and established an agricultural military colony (*Colonia Agricola Militar*) in 1930 named Álvaro Obregón who, after being a general during the revolution, later became president (Rubin 1997). It was in this town that people had risen up, two years before my arrival, against the establishment of industrial wind turbines on the Santa Teresa sand bar.

Charis would become a popular *cacique* in the *Istmo*, fighting to establish nine *ejidos* in the region (Smith 2009). Constantly negotiating for more autonomy against Oaxaca and Mexico City, he was responsible for some of the first schools,[4] hospitals and roads in Juchitán, maintaining an authoritative role as the *Istmo*'s principle *cacique* (Smith 2009). Following Partha Chatterjee (1993), Jeffery W. Rubin (1997, 45) contends that from 1934–1960 the *Istmo* had a 'domain of sovereignty' that animated the culture politics that were in constant conflict and negotiation to defend local initiatives and regional autonomy, while also attempting to manage state interventions in the region. Not without enemies, Charis still managed a popular power base until his death in 1964. No political force after Charis would unify the *Istmo* in the same way, and his image became an important political symbol fought over by the PRI and the COCEI in the 1970s (Monjardin 1993; Salman and Assies 2007).

Land tenure and development projects have long been contentious issues. Notable developments in the *Istmo* have been the trans-Isthmus railroad (1917), the Pan American Highway (1950), the Trans-Isthmus highway (1960), the electric plant (1960s), the rice-processing plant (1960s), the sugar mill (1960s and 70s), the Benito Juárez Dam (1958) and related irrigation land disputes (1960s), the pacific coast PEMEX refinery in Salina Cruz (1979), and later the development of industrial wind turbines with the completion of the first project Venta I in 1994 (Campbell et al. 1993; Hamister 2012). The planning, decision and construction of the Benito Juárez Dam was carried out in Mexico City and became intertwined in a series of land and resource management issues (Warman 1993), triggering conflict over private and social property—*ejidos* and communal land. 'Until the Benito Juárez Dam', writes Leigh Binford (1985, 184), 'there was little necessity for documenting one's [land] title in writing'. On 15 February 1960, the DAAC (*Departamento de Asuntos Agrarios y Colonización*) called for all the private property owners located in the *ejido* and communal lands of Juchitán to present proof of ownership in ninety days to begin the development of irrigation works. A series of land resolutions developed concerning the status of *ejidos*, communal or private lands, all to facilitate the construction of the irrigation system of the Benito Juárez Dam. Two resolutions in 1962 and 1964 became the centre of controversy, triggering a revolt from farmers and landowners who regarded these resolutions as acts to destroy existing land rights and relationships (CODIGODH 2014).

Warman (1993, 103) observes how 'communal lands were converted into *ejidos* by one decree, only to be converted into private property by a second decree, both expedited by Día Ordaz'. The situation was tense, as the state and other advocates among the *mestizo* and Zapotec elites illegally sought to transform social property into private property by paying taxes on communal lands in an attempt to make townships provide a paper trail to make a private claim on communal land (Warman 1993). On 31 March 1966 a resolution would be passed to normalize the privatization of 3,887 land titles, however, as noted by Binford (1985, 192), the titles 'only guaranteed the possession of lands of communal origin, not the *ownership* of those lands'. What emerged in the 1960s was a series of contradictory, ambiguous and error-filled land resolutions from the federal government, which failed to adequately address the land question, leading to the continuation of the conflict into the 1970s with the emergence of The Isthmus Coalition of Workers, Peasants and Students (*Coalición de Obreros, Campesinos y Estudiantes del Istmo)*, known as the COCEI.

The COCEI

The Isthmus Coalition of Workers, Peasants and Students or COCEI emerged from two principal forces taking place in the *Istmo*. The first was the Istmeños who had been involved in the 1968 student movement, experiencing and surviving the repression and Mexico's 'Dirty War' led by Gustavo Díaz Ordaz—the president behind the 1968 Olympic Tlatelolco massacre that killed (by conservative estimate) 200–300 people (Hodges and Gandy 2002). This event provided a political education, leading people towards grassroots organizing and militant forms of struggle (Rubin 1997). The second force was the timely *Istmeños* tensions toward autonomy that emerged from opposition to the PRI candidate Manuel Musalem (Tarú), who won elections in Juchitán in 1971 (Campbell et al. 1993). The Tarú administration did not last long, but it did inspire engagements with electoral politics as a site of struggle in the *Istmo* (Rubin 1997). The group that would later become known as COCEI carried out their first campaign in 1973 to force Dr. Barragán out of his chief administrative position at the public hospital. The schools required that all students have a health certificate, for which Barragán was illegally charging students. After a series of demonstrations, people took over the hospital and Barragán was removed (Rubin 1997).

The newly formed COCEI now turned their attention to the countryside. Starting in 1974, the COCEI held demonstrations against an Ejido Bank that denied *ejido* credits to small land owners and instead gave them to agriculturalists who rented smallholder land. They went on to help organize unions and demand the removal of military forces that were supporting timber companies in the Chimalapas. Important, was what later amounted to a successful land invasion in La Ventosa that began in 1974 to challenge private property titles distributed in 1966 by Díaz Ordaz. The COCEI filed lawsuits against private property titles in 1975–1977, expanding rural autonomy by shutting down the Communal Land Commission in 1978, while also helping the people of Álvaro Obregón to regain 1000 hectares of land they had been fighting for since 1964 from Federico Rasgado (Rubin 1997; Campbell 1994). Despite increasing repression by the state, elite and commercial interests in the area, COCEI activities were met with success.

The movement's success did not come without blood, sweat and disappearances. Juchitán Mayor Mario Bustillo and Oaxacan Governor Zárate Aquino would systematically send the police and military to beat, imprison and kill protesters who were blockading roads, occupying lands and buildings. In May 1974, COCEI leaders Daniel López Nelio and Oscar Matus were beaten by police and hired thugs by the then-treasurer of Juchitán. Later, Hector Sánchez was attacked in an electoral campaign, and that same month Lorenza Santiago was killed protesting electoral fraud in front of Bustillo's houses—the first *coceistas* to be murdered. The trend continued with two armed attacks in December, when a *priísta* attacked COCEI activists and the mayor's bodyguard attacked a group of small farmers (Rubin 1997, 140). This conflict in the *Istmo* intensified a year later at a demonstration marking Santiago's murder, when gunmen were paid to open fire: they killed seven farmers and one student. Again in 1977, the police attacked and jailed thirty-seven students. Rubin (1997, 142–43) notes that the '[a]rmy interventions did not generally result in reported injuries or deaths, while police activities frequently did'. Martínez López summarizes this experience:

> If for Juchitecos one is Juchiteco above all, and if one has roots in a strong sense of belonging to the community, then to feel massacred, persecuted, and beaten by soldiers and police creates a collective consciousness of solidarity. . . . Since the massacre of 1977, the *coceistas* have been surrounded by martyrs (Quoted in Rubin 1997, 43).

This cycle of political success and repression would continue into the 1980s. Then in 1981 Leopoldo de Gyves of the COCEI, in alliance with the Mexican Communist Party, won the municipal elections in Juchitán, becoming the first leftist electoral winner in Mexico (Rubin 1997; Altamirano-Jiménez 2014). In the 1980s the CO-CEI broadened their fight. They instituted the *Ayuntamiento Popular* (the people's government), repaired unpaved streets, built and staffed health clinics, established a public library in downtown Juchitán and rebuilt the dilapidated city hall (Rubin 1997, 166). They also supported workers in labor struggles against Coca-Cola and a beer bottling plant, negotiated with state and federal authorities to secure agricultural credit and crop-insurance settlements for peasant farmers, and supported a land occupation in Colonia Rodrigo Carrasco (Rubin 1997). These actions came with political consequences, as both the state and federal agencies denied loans and credit to the municipal government, businesses held strikes against the COCEI, and a right-wing paramilitary group known as the Committee for the Defense of the Rights of the People of Juchitán began carrying out armed attacks against COCEI supporters (Campbell et al. 1993; Rubin 1997). The reaction against the COCEI began to move in the direction of civil war politics akin to those of Central America at the time, albeit less extreme. Against the backdrop of booming oil production in Salina Cruz, another violent incident, this time in 1983, provided the federal government with the pretext to invade Juchitán with military forces, depose the COCEI city government, repress COCEI activities, disappear Victor Pineda Henestrosa—Victor Yodo—and set up barracks around the city during a five-year occupation of the city hall (Campbell et al. 1993; Rubin 1997). In 1986 the COCEI returned to power through a strategic coalition with the local PRI, causing discord within the COCEI and their followers (Altamirano-Jiménez 2014; Rubin 1997). Support for the group continued despite disagreement and concern.

The COCEI was at the forefront of radical leftist politics in the 1970s and early 80s in the way they combined Marxian ideologies (Leninist and Maoist) with Zapotec culture. This was a major political strength of the COCEI. They rooted their activities in Zapotec identity, brought Zapotec language into schools and made it the dominant language of the region. Zapotec attire and footwear would become important political symbols for politicians and their constituency (Campbell et al. 1993). The COCEI added another chapter to the insurrectionary history of the *Istmo*, articulating a cultural politics that institutionally revitalized and created space for Zapotec and other Indigenous cultures that have faced subjugation and compromises with invading forces for 500 years. The group underwent significant changes, however, after the repression of 1983.

Arguably the leftist politics began to subsume Zapotec indigeneity and land relations. Rubin (1997, 133–34) comments on the COCEI's turn to negotiation with the Mexican state by writing:

> Because leftist parties were illegal for most of the 1970s, COCEI acted independently of political parties by necessity as well as conviction. In this way, the exclusion of the left from electoral competition reinforced the independent character of grassroots opposition. Successful grassroots organizing in the 1970s, in turn, prompted the regime to attempt to control leftist political activity by channeling it into political parties and electoral campaigning in the 1980s.

It was thought that COCEI's militancy, cultural appreciation and opposition to the politics of modernization would keep it from falling into the traps of the Mexican political system (Campbell et al. 1993; Rubin 1997). Signs of this began to faintly emerge in 1986. The new collaboration with the PRI resulting from the past decade of negotiation and repression signalled a new political formation. This caused discord amongst the COCEI leaders and supporters, but the collaboration continued. Shortly after reelection in 1986, the COCEI started to downplay its Zapotec identity in an attempt to gain a wider political following (Altamirano-Jiménez 2014). The COCEI's leftism was starting to outweigh its Indigenous politics.

This channelling of COCEI activity became apparent in 1989, during Hector Sánchez's administration (1989–1992). Aside from the first corruption accusation to be waged against the COCEI administration, Sánchez's negotiation with Carlos Salinas de Gortari became a major turning point as both the Party of the Democratic Revolution (PRD) and COCEI openly opposed Gortari. Coming to power on the back of a fraudulent victory later ratified by the PRI House of Deputies, Gortari would become infamous for instituting structural adjustment programs (SAPs) that would eliminate trade barriers, privatize large sections of the public sector and significantly cut social welfare programmes. Gortari's legacy would be that of the president who signed NAFTA and triggered the Zapatista uprising in 1994 (Rubin 1997; Stephen 2002). This economic restructuring opened the Mexican economy to new forms of biotechnology[5] and intellectual property rights, which combined with dramatic social spending cuts to agricultural subsidies in 1991, ushered in a potentially dramatic transformation. Crop insurance was abolished, farmers had to work with commercial banks, and fixed prices were eliminated for corn and beans that would later combine to displace some 15 million small-holder farmers (Assies 2008; Bello 2009, 49). Gortari's administration amended Article 27 of the Mexican constitution to allow social property—*ejidos* and communal land—to be formally titled, creating the possibility of privatization. This allowed a change in the use of soil so that development activities other than agriculture could be performed on *ejido* and communal land (Assies 2008; Stephen 2013)—enabling, among other things, wind turbine development.

Sánchez's choice to allow Gortari was strategic. Campbell (1993) shows that accommodating Gortari reaped many benefits: (1) a relationship with the Gortari administration could help undermine the PRI in the 1992 elections and most important (2) the COCEI would have access to federal funds. This choice resulted in dividing COCEI members into reformist and radical factions. Discussing this incident with a COCEI contemporary of Sánchez, Dona explained:

> **Dona:** The COCEI was divided by the government, one by one. One by one, individually through *compadrazgo's*,[6] money, goods (land, tractors, etc.). . . .
>
> **AD (Alexander Dunlap):** *The way you told it was that COCEI sold out when they let Salinas de Gortari come to town?*
>
> **Dona:** One of them, not all of them. Daniel [Nelio] was pissed and everyone else too was freaking out—it was just Hector. It wasn't the COCEI, everybody else was furious. Then they [the government] paid them off and everything calmed down.

AD: *At the time of Hector selling out, was it the majority that was outraged?*

Dona: Many of the COCEI leaders were bothered and upset because they were putting their ideals to the side, but most of the town benefited. There were streets paved, goods, and services—so really he was not criticized a lot, because people really liked the changes.

AD: *They liked development?*

Dona: Exactly.

AD: *From your perspective now, do you think his selling out was good?*

Dona: That day I was so angry I could have died. To me it was horrible. I did not want to betray [Cuauhtémoc] Cárdenas, [7] I followed Cárdenas word by word. He [Hector] as a human being justifies himself, I do not agree with him, but oh well. It is not like I am criticizing him all the time, but occasionally I will confront him about it, but he justifies himself and a lot of people tell me: "'He did not do anything wrong. He brought social benefits'." A lot of people told me, "'Hey, he fixed up city hall, he fixed up the streets'," because things were in bad shape—terribly bad. Things were really backwards. He really changed everything.

While Gortari's visit was met with protest, Sánchez took political responsibility for it, signing the *concertación* accords (or the *Pacto de Concertación Social*) (Rubin 1997). This brought federal funds from the National Solidarity Program (PRONOSOL) that offered economic support to indigenous groups if they would stop blockading roads and protesting and embrace instead the (neoliberal) social peace being established by Gortari (Altamirano-Jiménez 2014). With this money, Sánchez provided more public sector jobs than any other mayor, paving streets, installing drainage, improving local infrastructure and developing Juchitán as it had never been developed before (Rubin 1997). However, this leaves us to wonder if the COCEI could have served to amplify the Zapatista uprising if they had continued to reject Gortari and his neoliberal programme, serving as a front against the continued neoliberalization of the Mexican state.

Wind Turbines

Since the *concertación* accords, the COCEI, sometimes in alliance with the PRI, PRD and PT (Labor Party), 'has slowly "welcomed" supermarkets, burger chains, and wind parks', writes Isabel Altamirano-Jiménez (2014, 196). Planning for the first wind turbine pilot project began as early as the late 1980s, and this was completed in 1994. La Venta I, consisting of seven industrial wind turbines, was financed by the Mexican government and the World Bank and was initially constructed to assess the energy potential as well as human and ecological impact, but, as many locals will contend, the assessment never took place (Friede 2016; Simon 2013; Hamister 2012). Wind development dramatically increased in the Isthmus after the 2003 report by the National Renewable Energy laboratory[8] titled *Wind Energy Resource*

Atlas of Oaxaca (Elliott et al. 2003), which proclaimed that the *Istmo* had 'excellent' wind resources. Around the same time the COCEI were reluctant to take a clear position against the Central American Plan Pueblo Panama (2001), later called the Mesoamerica project, which was an enormous multi-country integrated mega-infrastructure development project (Altamirano-Jiménez 2014). Concerns about the COCEI were amplified three years later when the first Walmart was opened in Juchitán in 2004 (Call 2011). While many were enchanted by the 146 permanent jobs and the Jimador tequila, many others saw the store as an assault on Juchitán's long-prided Zapotec artisans, small businesses and farmers (Call 2011), making this and the later wind turbines the final straw for many COCEI supporters.

The way the wind companies arrived deserves attention and detail. As eloquently summarized by a member of the Tepeyac Center for Human Rights (CDHT):

> The politicians of that time knew how important this whole part [of the *Istmo*] was, without touching Tehuantepec or Salina Cruz and all of the area around Juchitán and Ixtepec they began talking about green energy, it was a project that had long-term potential. With the changing of administrations, all of these lands were brought into line. The politicians and the rich who had access to this information bought up a lot of land. In the *Istmo* the wind energy project became possible because the Left permitted it. We are not just talking about the PRI, we are talking about the COCEI—they allowed it. Because it was a novelty—alternative energy, green energy—there was no problem, it was the most revolutionary thing. It is as if we were talking about independent media [community radio], we will choose our own programming, it will not be for profit and they [the people] thought the same way about green energy—it was clean energy, it didn't pollute and it was seen as a proposal against the system.
>
> As the companies began to arrive, the government was not even acting as the middle man; they came directly to the communities and with specific people in those communities. They went directly to those people to start dealing, when we started to realize what was going on we were seeing wind projects being built in this part of the Isthmus [pointing to the north, near La Venta]. The modernistic inertia was advancing until the notion of wind energy fractured the COCEI itself. So there was no doubt that wind energy was the least damaging of the ways to produce energy—it's better than a refinery, better than a thermal electric plant, it's better than a hydroelectric plant and the argument that these were lands not being used started to spread. And where the fracture began was when one of the local politicians started to discover how these people [the companies] were coming in was not very clear. They were territorializing the region, they were taking over the land for thirty or forty years without allowing people to use their land as they were accustomed. That was an aggression right there. This is part of the history with the Zapotec people, so here we started to see outbreaks of protests in places like Unión Hidalgo, in Juchitán, in La Venta, in Santo Domingo [Ingenio], but the protests were, well ... a large group of people did not want to sell, but the majority of people were protesting because they wanted to be paid more for their land. So this is part of the history, the history at this point divided in two. The Zapotecs [in the Northern coastal *Istmo*] allowed them to fill their lands with wind turbines and there was some institutionalized violence, some people were repressed. The government contacted the leaders and those they could not buy off, they repressed.

The other half of the history is the Ikoots and Zapotecs in the South, who began organizing in the late 2000s to resist and militantly reject wind energy projects spreading into their territory. Resistance against wind energy development was met with violent repression from authorities and, under politico-economic pressure, the struggle turned into a protracted inter-communal conflict that will be examined and discussed at length in Chapters 2–5.

While the wind rush was taking hold in the *Istmo*, political turbulence and instability engulfed Oaxaca City. In May 2006 for the 25th consecutive year, the Section 22 teachers union commenced their strike and descended on Oaxaca City's *Zócalo* demanding improved educational facilities and pay (Vásquez 2007). On 14 June, strikers were confronted by over 3,000 heavily-equipped state police who came to clear the *Zócalo*, which resulted in a protracted street battle with police leaving hundreds injured and hospitalized. This police action, ordered by Governor Ernesto Ulises Ruiz Ortiz, aggravated the situation, triggering widespread mobilization and unification of various social sectors to demand the resignation of Governor Ortiz (Gibler 2009). Municipalities, unions, non-governmental organizations, civil society groups, cooperatives and others came to form the Popular Assembly of the Peoples of Oaxaca (*Asamblea Popular de los Pueblos de Oaxaca*) or APPO. Days later, on 17 June, APPO reconvened in the *Zócalo* to establish an occupation and declared itself the governing body of Oaxaca, consequently plunging the city into a state of permanent protest and rebellion against the governor. Barricades were erected across the city to defend what turned into a commune from the police and paramilitary raids. Seeking country-wide solidarity, APPO began organizing popular assemblies at various levels—neighbourhoods, street blocks, unions and towns—and was met with repression from police and plain-clothed paramilitaries. Paramilitary units were driving around in pickup trucks, often masked, with automatic and high-powered rifles that dispensed terror and became known as the *caravanas de la muerte* (The Caravans of Death) (see Stephen 2013; Vásquez 2007). The Oaxaca Commune lasted until December 2006 with little resolution, resulting in disappearances and deaths, some say twenty-seven (Denham 2008; Vásquez 2007), which attracted condemnation from human rights groups concerning the Mexican governments use of death squads, summary executions and violating the Geneva Conventions standards that prohibit attacking and shooting at unarmed medics (Gibler 2009; Stephen 2013). This history of upheaval and repression, we will learn in Chapter 3, is intertwined with actors promoting wind energy development and seeking to undermine the resistance against these projects in the *Istmo*.

Since the publication of the *Wind Energy Resource Atlas of Oaxaca (2003)* and the Oaxaca Insurrection (2006), wind energy development in the *Istmo* has skyrocketed. Wind energy extraction is contracted in the *Istmo* through three principal methods. First, and the minority in the *Istmo*, are wind parks financed by the public sector and operated by the Federal Electricity Commission (CFE). Second, Independent Power Production (IPP) are wind parks built and operated by private companies, yet the energy generated from them is sold exclusively to the CFE (Juárez-Hernández and León 2014). Third, 'self-supply' (autoabastecimiento) is when the electricity extracted and produced by wind energy projects is held privately, reserved for the investors or co-owners of the project (see Chapter 2). According to Sergio Juárez-Hernández and Gabriel León (2014, 4), '[m]ore than 90% of the [energy] capacity belongs to private companies, of which two, Acciona Energía and Iberdrola, account for nearly 65% of the regional share.' The *Istmo* is the largest area of wind energy development site in Latin America and accounts for about ninety percent of Mexico's renewable energy, with more planned (Rivas 2015). Following the estimate of Rivas (2015), the *Istmo* is recorded having 1,642 wind turbines built, which includes the potential to generate 10,000 megawatts with the possibility of building over 5,000 wind turbines (Bessi and Navarro 2014; Rivas 2015; Rubí 2016). Affirming this trajectory of wind energy development, Oaxacan governor Gabino Cue Monteagudo would continue to claim 'that in 2016 Oaxaca will have 23 more wind farms' (Rivas 2015, A20).

What constitutes a 'wind park' is largely undefined, and accounting for the exact number of wind turbines built in the *Istmo* is difficult due to their rapid development, various phases of a single project and how the names of wind companies and parks are used interchangeably in different sources. To give an idea of the extent of wind turbine development (see Table 1.1), the parks are reviewed by year, which is as complete as possible based on triangulating wind parks through different sources (see ADMEE 2010; Oceransky 2011; CDM 2012; Hamister 2012; SIPAZ 2013; Simon 2013; Mejía 2014; Juárez-Hernández and León 2014). The wind parks emerging after La Venta I in 1994 have significantly larger turbine towers of forty-five to ninety metres (approx. 148–295 feet) in height, with blades between ninety and 110 metres (approx. 295–360 feet) in diameter depending on the project and wind turbine. La Venta II (83.3 MW) was constructed in 2007, which was followed in 2009 by three wind parks: Parques Ecológicos de México (80 MW), Instituto de Investigaciones eléctricas (5 MW) and Eurus (167 turbines/ 250 MW). In 2010 there would be two industrial wind parks built: the Eurus II (212.5 MW) and La Mata-La Ventosa (67.5 MW). The following year a set of four parks called Oaxaca I, II, III, and IV were completed, generating a total output of 405.60 megawatts. In 2012 there would be six wind parks constructed: Bii Nee Stipa Enería Eólica I (26.35 MW); Bii Nee Stipa Enería Eólica II (288 MW); Zopiloapan- Bii Nee Stipa III (70MW); La Venta III (101 MW); Piedra Larga phase I (90 MW); and Fuerza Eólica del Istmo phase I and II (180 MW).

In 2013 another six wind parks were built with Piedra Larga phase II (137.5 MW), Bií Stinú (164MW/87); Eoliatec del Istmo (226 MW); Energía Alterna Istmeña (250 MW); Oaxaca Eoliatec del Istmo phase II (70MW) and Eoliatec del Pacifico (215.6 MW). Then in 2014, the Santo Domingo (160 MW) wind energy park was completed, which was followed by 117 turbine Bíi Hioxo wind park with a capacity of 234 megawatts (see Chapter 3), which was the first park built next to the Laguna Superior. Finally, as of 2 March 2016, a new thirty-four wind turbine park Sureste (Southeast) I phase II was opened in Asunción Ixtaltepec (ADNsureste, 2016). This list (and corresponding Table 1.1) identifies wind parks and their phases by

Table 1.1. Wind Parks in the Isthmus of Tehuantepec by Year.

	Name	Year	Energy Production Regime	Megawatt Capacity	Turbine #
1	La Venta I	1994	Public Sector	1.6	5
2	La Venta II	2007	Public Sector	83.3	41*
3	Parques Ecológicos de México	2009	Self-Supply	80	40*
4	Instituto de Investigaciones Eléctricas	2009	Independent Power Production	5	3*
5	Eurus	2009	Self-Supply	250	167
6	Eurus II	2010	Self-Supply	212.5	106*
7	La Mata-La Ventosa	2010	Self-Supply	67.5	34*
8	Oaxaca I	2011	Independent Power Production	102	52*
9	Oaxaca II	2011	Independent Power Production	102	50*
10	Oaxaca III	2011	Independent Power Production	102	53*
11	Oaxaca IV	2011	Independent Power Production	102	50*
12	Bii Nee Stipa Enería Eólica I	2012	Self-Supply	26.35	13*
13	Bii Nee Stipa Enería Eólica II	2012	Self-Supply	288	144*
14	Zopiloapan- Bii Nee Stipa III	2012	Self-Supply	70	35*
15	La Venta III	2012	Independent Power Production	101	50*
16	Piedra Larga Fase I	2012	Self-Supply	90	45*
17	Fuerza Eólica del Istmo Fase I & II	2012	Self-Supply	180	90*
18	Piedra Larga 2ª Fase	2013	Self-Supply	137.5	69*
19	Bíí Stinú	2013	Self-Supply	164	87
20	Eoliatec del Istmo	2013	Self-Supply	226.8	113*
21	Energía Alterna Istmeña	2013	Self-Supply	250.9	108*
22	Oaxaca Eoliatec del Istmo Fase II	2013	Self-Supply	70	35*
23	Eoliatec del Pacifico	2013	Self-Supply	215.6	107*
24	El parque eólico Santo Domingo	2014	Self-Supply	160	80
25	Bíí Hioxo	2014	Self-Supply	234	117
26	Sureste I	2016	Independent Power Production	68*	34
					Total: 1,728

SOURCE: AUTHOR

year, along with the type of energy generation regime, megawatt capacity and number of turbines. The wind parks were triangulated through a variety of sources up to the winter of 2016 (see ADMEE 2010; Oceransky 2011; CDM 2012; Hamister 2012; SIPAZ 2013; Simon 2013; Mejía 2014; Juárez-Hernández and León 2014). The asterisks (*) identify estimations[9] of wind turbines (or megawatt production) that accounts for a total of 1,728 wind turbines. This book relies on the more conservative estimate of 1,642 turbines mentioned above, although the likelihood of wind turbines surpassing 1,728 is entirely plausible. The Mareña Renovables/Eólica del Sur wind park is not listed here (see Chapters 4 and 5) because project construction has not begun. Future wind parks are planned and are being negotiated.

CONCLUSION

This chapter offered a broad genealogy of the *Istmo* from pre-colonial times to the emergence of wind turbines. Recognized for its colonial adaptation and negation, the *Istmo* maintained a high propensity for rebellion and insurrection against external authorities deemed unfair or tyrannical, whether it was the Spanish, French or interventions from the Mexican state. *Istmeños* have a reputation instigating insurrection, which coalesces into the general desire for the autonomy and self-determination of the region. While the land and natural resources of the *Istmo* was a central concern during the colonial period, the struggles over land, specifically communal land, have emerged as increasingly important issues with the advent and consolidation of the Mexican state. The issue of communal land, and its land titling and regularization, took centre stage in the 1950s and 1960s and remains highly salient (as will be discussed in the following chapters). The continuity of the struggle over cultural autonomy, dignity and natural resources stretches back long before the Mexican Revolution and continues today with the arrival of 'wind rush' in the region.

The next chapter begins by examining life surrounded by wind turbines in La Ventosa. This is a town located in Juchitán County, a twenty-five-minute bus ride north of Juchitán, where (by my accounting) about eighty to ninety-five percent of the town is surrounded by layers upon layers of wind turbines and electrical infrastructure. This chapter serves as a starting point to get an idea about life enclosed by wind turbines, which provides a foundation for later Chapters (3 and 4) that document the struggles against wind energy development. Before moving to the next chapter, the words of Adriana López Monjardin (1993, 69) when discussing the cultural debates in the Istmo remain insightful: 'the only really "traditional" aspect of the Juchitecos is their rebelliousness in the face of oppression and their flexibility in adapting to change, while conserving and refining their specificity'. This tradition is undoubtedly alive and well in the struggle against wind energy development, even if in later chapters we will focus on the structural oppression faced by communities who are forced to negotiate, adapt and reappropriate for their survival and advancement.

NOTES

1. Smith (2009, 202) goes with the Agrarian Department figures, even if these are lower, but believes this is because of double titling and making fake land titles.

2. *Programa de Certificación de Derechos Ejidales y Titulación de Solares Urbanos*/Program for the Certification of Ejido Land Rights and the tiling of Urban Housing Plots.

3. See notes on Pistoleros and Popular movements (Smith 2009).

4. Zapotec Charis used his inadequate Spanish skills as a way to negotiate a school, using discrimination against Zapotecs as means to broker development (Rubin 1997, 50).

5. Notably Monsanto's notorious "'Terminator' "Seeds, which would die after one growth cycle (See Bello 2009).

6. This word references the reciprocal social relationship that exists between godparents and godchildren: http://www.meriam-webster.com/dictionary/compadrazgo.

7. The 1988 presidential candidate competing with Salina de Gortari and founder of the Party of the Democratic Revolution (PRD) who was in alliance with the COCEI at the time.

8. This report was funded by US Agency for International Development (USAID); Secretaría de Energia (SENER); Gobierno del estado de Oaxaca. Secretaria de Desarrollo Industrial y Comercial (SEDIC); Comision Federal de Electricidad (CFE), Instituto de Investigaciones Electricas; Comision Nacional del Agua (CONAGUA) and Truewind Solutions, LLC.

9. The estimations are assuming that roughly each turbine generates 2MW, as is common. Many wind turbines only generate 1.5MW or up to 3MW. The latter is more common for newer generation wind turbines which are increasing their megawatt output with technological innovations.

2

'We are surrounded': Living under Wind Turbines in La Ventosa

> Not everything that shines is gold.
> —Etelvina Valdivieso

La Ventosa—which means 'the windy place'—sits northeast of Juchitán on a savannah plain with scattered hills, near the base of the Sierra Atravesada mountain range, with winds sweeping into the town and south across the coastal *Istmo* all the way to the Pacific Ocean. Winds of 200 kilometres per hour have been recorded in La Ventosa, uprooting palm trees, throwing horses to the ground and flipping semi-trailer trucks on their sides (Campbell 1994). La Ventosa was the second town, after La Venta, to begin wind energy development, and the second town to become surrounded by wind turbines—and now appears to be eighty to ninety-five percent enclosed by wind turbines.

To better understand the experience of living under and surrounded by wind turbines in La Ventosa, this chapter draws on secondary research material, participant observation and sixty-three semi-structured personal encounters and interviews. Most alarming here are the concerns regarding the intense environmental impact on the land as well as human health, with local doctors claiming that that the majority of people in the town have some type of cancer—with incidences apparently increasing since the introduction of the wind parks. This section is followed by an examination of the benefit sharing and social development emerging from wind energy projects, raising issues concerning the collective benefits for people in La Ventosa and the emergence of a type of rural gentrification that continues to have deleterious effects on the town. This chapter demonstrates how wind turbine development in La Ventosa has intensified pre-existing patterns of land use, inequality and poverty in the town.

WIND PARKS: CONSTRUCTION, ENVIRONMENTAL IMPACT AND FINANCE

Wind park planning in La Ventosa began in the late 1980s when the first commercial project—the Parques Ecológicos de México (80 MW) was built by the Spanish companies Iberdrola and Gamesa—and was completed in 2009. Development continued through 2010 with the completion of two wind parks: Gamesa's Bii Nee Stipa I-III ('El Retiro') (74 MW) by Gamesa and the La Mata and La Ventosa Wind Park (65.7 MW) by Électricité de France (EDF). Three parks were completed in 2012: Fuerza Eólica del Istmo 1a (30 MW), 2a (50 MW) by General Electric Wind and Stipa Naaya (74 MW) park by Gamesa (Mejía, 2014). The latter three projects extend the field of wind parks around La Ventosa to the towns of La Mata, Espinal and Juchitán. On 30 July 2015, a Free, Prior and Informed Consent (FPIC) consultation approved the Eólica del Sur wind park on the outskirts of Juchitán (see Chapter 5), which if completed would concentrate wind turbines into the last two remaining sections of land without wind turbines, between Juchitán and La Ventosa. The arrival of the FPIC might serve as an example of the Mexican state's placing 'an occasional brake' on controversial development projects in order to preserve political legitimacy in the area (Borras et al. 2012, 411; Wolford et al. 2013), yet as we will see in Chapter 5 there is more to the orchestration of the FPIC process.

Cost, Finance and Greening Industrial Development

Wind energy projects require substantial infrastructure investment. The International Renewable Energy Agency (IRENA) estimates that a twenty megawatt onshore wind farm costs a total of $44.7 million (IRENA 2015, 58). Also, it is said that a single wind turbine can cost $1.3 to $2.2 million per MW (Windustry 2012). Construction costs depend on the local political, economic and environmental context, however, and the high capital expenditures mix with the (previously existing) perception of low-reliability and profitability of wind energy that causes doubts among investors. These doubts were eased over time as economic liberalization and climate change legislation[1] promoted foreign direct investment (FDI) in Mexico, with wind energy representing one of the latest investment frontiers and part of the global 'Green Rush' for renewable energy—the new gold.

Mexico received thirteen structural adjustment loans from the World Bank from 1980 to 1991, followed by three notable policies that laid the foundation for the Oaxacan wind rush. The first was the revision of Article 27 in 1992 that created the possibility of privatizing social property—*ejidos* and communal land (Chapter 1) The second was the 1992 electricity law that contains a notion of 'self-supply' (*autoabastecimiento*) allowing transnational corporations to manage and build their own energy production facilities, enabling shareholders to decide on how and where the energy produced would be used with any excess electricity generation being sold back to the national grid managed by the Comisión Federal de Electricidad (CFE)

(USAID 2009). In this particular case, the self-supply mechanism directly allows the privatization of wind resources through privately managed wind parks that make up 'more than 90%' of the wind energy generated in the region (Juárez-Hernández and León 2014, 3). Third was the notorious 1994 North American Free Trade Agreement (NAFTA), which states in Section 6 that 'an enterprise may acquire, establish, and/or operate an electrical generating facility in Mexico to meet the enterprise's own supply needs' (USAID 2009, 2). This made the 1992 electrify law transnational, opening up Mexico to foreign corporations. Now President Peña Nieto took the process even further, with his 2013 Petroleum and Federal Electric Utility Act mandating that social property or other land holders must negotiate and eventually surrender their land to energy companies in regions of development interest (Payan and Correa-Cabrera 2014). This legislation privatized the two largest firms in Mexico: PEMEX and the CFE. This new privatization law was forced through after decades of struggle and repression by various segments of the population resisting privatization, among them the Mexican Electrical Workers Union (SME) (González 2014). Regarding this legislation, Cuauhtémoc Cárdenas went so far to declare: 'Never throughout our history as an independent nation has the country seen such a dismantlement of the protections to our sovereignty and self-determination' (Cypher 2014, 27).

The Mexican government created openings to establish a profitable renewable energy market, gaining the attention of Danish (Vestas), Spanish (Iberdrola, Gamesa, Marña Revonables/Eólica del Sur) and United States (Clipper) wind companies also gave rise to the establishment of a series of limited liability companies in Mexico organized to pursue large-scale wind energy exploitation. In the case of La Mata and La Ventosa Wind Park, Eléctrica del Valle de México (EVM) understood that Walmart wanted to 'go green,' approached them, and negotiated a sixty percent power share of the La Mata and La Ventosa wind park with electricity bought 'at a price that is higher than wholesale, but lower than retail' (WDM 2011; USAID 2009, 5). Other investors in the wind parks around La Ventosa are Cemex and Grupo Bimbo—respectively the world's largest cement and food processing corporations—and two of Latin America's largest mineral extraction and processing companies: Grupo Mexico and Peñoles (Garcia 2012). Grupo Mexico has thirty-seven turbines in phases II–III of Bii Nee Stipa Wind Park, proving 'once again', in their words, their 'commitment to sustainability and the environment' (GrupoMexico 2014, 7; Hristova 2014).

Also demonstrating their 'commitment to sustainability and environmental stewardship,' Peñoles (2014, 58) has two self-supply wind parks: Fuerza Eólica del Istmo 1a and 2a around La Ventosa, which includes a gold mine concession of 10,039 hectares with the Canadian River Resources Inc./Arcus Development Group just inside the town (Chaca 2015; Biller 2012; RS 2008). Initially, Peñoles was a principal sponsor in the recently denied Eólica del Sur wind park (AMDEE 2012), but due to delays investments have now headed north into an 180MW wind park in Coahuila (Peñoles 2014). These wind parks are not only built on alliances between state, national and large-scale foreign capital, but also operate in collaboration with other industries that are justified with the green economic logic of 'offsetting' (Borras et al.

2012; Holmes 2014; Böhm and Dabhi 2009; Sullivan 2009; 2013a; 2014; 2017a). 'Offsetting' environmental damages has a history with mining companies that attempts to rebrand mining, and all of its negative natural environmental connotations, as 'sustainable mining' (Kirsch 2010). Mining companies such as Rio Tinto are responsible for designing environmental impact offsetting policy, reducing the intricate and context specific attribute of ecosystems to justify resource extraction (Seagle 2012). This extends to creating concepts such as 'green uranium' (Sullivan 2013b, 82), enabling resource extraction companies to continue, if not intensify, their environmentally destructive operations with their newfound public relations propaganda package used to potentially defuse, fragment and mitigate popular resistance against these projects in both the Global North and South (Dunlap and Fairhead 2014). In La Ventosa, wind turbines have now been integrated into this process of greening environmental degradation and renewing environmental destruction associated with extraction and processing industries, propelling industrial growth forward with new possibilities of receiving 'climate' and 'clean' technology funds and loans.

The La Mata and La Ventosa Wind Park became the 'poster boys' for the World Bank's leading Clean Technology Fund (CTF) project in Mexico (WDM 2011). The CTF is the largest of the World Bank's Climate Investment Funds (CIFs), which were designed to support low-carbon technologies and encourage 'clean' investments. The World Bank also invested in La Venta II with a USD 12.9 million loan and La Venta III with USD 25 million grant, both about thirty minutes' drive from La Ventosa. La Mata and La Ventosa received loans from the CTF (USD 15 million concessional loan); International Finance Corporation (IFC) (USD 23.68 million); the Inter-American Development Bank (IDB) (USD 21 million) and Export-Import Bank (USD 80 million) (WDM 2011). This wind park was registered with the Clean Development Mechanisms (CDM), which is projected to generate 1,179 certified emissions reduction credits (CERs) over the next seven years and which have been 'forward sold' to EDF Trading giving it the possibility to gain over USD 40 million from the CDM (CDM 2006; WDM 2011, 11). Despite problems with democratic participation and oversight with the CDM (Newell 2014), it has helped reduce the risk to investors, realizing the profitability of wind energy generation and renewing the prospects of wind energy development for transnational corporations.

Notably, following the self-supply status of the majority of the wind parks in this region, the energy generated from these wind projects are exported to industrial areas in Mexico, Guatemala or the United States (WDM 2011; USAID 2009). Rumours have spread about towns negotiating free electricity, and that the COCEI have been negotiating with the wind companies to have wind turbines to power Juchitán at a free or substantially reduced energy price, but the wind energy development model already in place is organized around an energy export-oriented model that seeks to power industrial zones, private industry and other countries. The CFE signed wind energy export agreements with Belize, Guatemala and Los Angeles, California as early as 2009 (USAID 2009). Similarly, the Mexican state and private developers have been trying to turn the *Istmo* into a *maquiladora* corridor, which was conceived as early as 2001 under Plan Puebla Panama with its vision carried forward in drug war legislation (2008 Mérida initiative), and the Alliance for Prosperity and Peace, which calls for expanding the electricity supply grid in Central America (Paley 2015).

Economic restructuring and neoliberalism mixed with concerns around climate change have articulated a hybridized style of neoliberalism that has supported the Global growth and expansion of the renewable energy industry. From a business perspective, transnational corporations have made the best possible use of existing national legislation and international programmes to 'roll out' new laws for businesses to grow and to develop renewable energy markets and annual revenues, while simultaneously demonstrating corporate attempts at social responsibility. The green economy can renew not only the public images of large transnational companies such as Walmart, Peñoles and Grupo Mexico but also stimulate revenue streams, creating the possibility for continued economic and infrastructural expansion. From the perspective of socio-environmental harmony and quality of life, these projects may just be reinforcing path dependency, expanding industrial consumption and growth. This deployment of industrial-scale wind energy is forfeiting alternative trajectories while promoting destructive environmental interventions, spreading high-consumption habits and significantly altering people's lives.

Figure 2.1. La Ventosa, Mexico. SOURCE: AUTHOR

THE POLITICS OF LAND ACCESS AND THE ARRIVAL OF WIND ENERGY

Toward the end of our first day going door-to-door in La Ventosa to ask people about the wind parks, my friend and I, with our local escort, dragged ourselves along, under the radiant sun through the San Miguel neighborhood. We came across a partially-completed foundation for a high-tension wire tower. The town had already been engulfed by industrial wind turbines and electrical infrastructure, so I was curious as to why they did not complete its construction. I asked our escort, and he explained. At first, without saying a word to anyone in the neighborhood, the state had decided to have high-tension wires built running through the neighborhood along the canal. Soon people began to ask the construction workers, 'what is that thing? What are you building?' The contractors told them: 'We are going to build power lines down here', pointing across the neighborhood. Immediately, the people started yelling, organized and came to seize the construction equipment and cars belonging to the company, forcing them to stop working, and eventually discontinue the entire project, leaving only a lonely concrete foundation. Nobody talked to the residents about the construction. The state did not consult them or ask anyone in that neighborhood, let alone the town. The people learned more by asking the construction workers, and through their own will terminated the construction. If people had not asked or pursued it further, they would have never known what was going on and would have been draped in another layer of electrical infrastructure—in this case high-tension wires. After hearing this story and looking around while standing in a neighbourhood at the centre of a wind-energy generation site, it became abundantly clear that this high-tension wire foundation was a microcosm of the development in this area: the people were never fully or properly informed and could only stop the construction on their own. Without action on the part of the towns' people, the land and neighborhoods are consumed by electrical infrastructural development and the town becomes engulfed by wind turbines.

Earlier that day a woman warned us: 'A lot of people are afraid to talk about the wind turbines because of the *cacique*'. The *cacique* is a political boss in a town or region, who typically maintains power through clientele relationships, making use of electoral fraud, intimidation and violent acts (Rubin 1997). In La Ventosa, as well as other towns in the *Istmo*, the *caciques* and their networks of state institutions, politicians, union leaders, land owners, business elite and gunmen make them central to any business and development project in a region. The caciques' control and influence on local politics and business makes them central players in wind turbine development, providing that bridge between the international and the local in La Ventosa. This struggle against regional political bosses is old and quite familiar. The notion of the *cacique* received contemporary significance after the revolution, which changed regionally with the death of *Istmo*'s populist *cacique* General Charis in 1964 (see Chapter 1). Charis' death opened a power vacuum for local and foreign elites in the *Istmo*. This intense and bloody struggle for regional power after Charis became a fierce and contentious battle between the PRI and the COCEI.

Agrarian politics and development has always been a struggle in La Ventosa. Back in the 1970s, an oral history interview explains that the current *cacique* in La Ventosa called on state forces to evict COCEI activists who were occupying a piece of communal land to protest its privatization. Today this same land is where the *cacique* resides. In further conversations and interviews, I learned that the *cacique* was involved in regional development and land change projects throughout the years. The *cacique*, along with other PRI politicians (like Wala Tina) were said to have 'destroyed the sugarcane refinery' and 'finished off the sugarcane business in Juchitán'. Dona, bringing it to the present, continued that '[t]he *cacique* has always had a lot of power, he had the PRI government with him and it is still like that [today]'. One land owner, Caballo, described how 'the politicians from here and all over the place got rid of sugarcane refineries and sugarcane was the best crop for La Ventosa'. Sugarcane was replaced with cattle rearing in 1989–1991 because, in the words of Caballo, 'the wind does not affect the grass much. It does affect it somewhat, the grass cannot grow a lot because when you water it the wind comes and takes it away and the grass gets hard. That is why the sugar cane was better, because it could withstand the wind'. These agrarian changes coincided with Mexico's economic restructuring and the COCEI's cooperation with the PRI in regional elections in 1986, as well as with Salinas de Gortari visit to the *Istmo* in 1989 (see Chapter 1). The changes took place against the backdrop of intense social conflict, death and people struggling for their lives after the 1983 military occupation of Juchitán and the more general effects of Mexico's neoliberalization (Stephen 2002; Bello 2009).

These macroeconomic, agrarian and political changes in La Ventosa set the stage for wind energy development. The local *cacique* led the way: 'He takes control over everything, he takes advantage of the situation before it is put into place,' said Caballo, who continued by recounting the arrival of the wind turbines:

> He made an announcement at the beginning: "anyone who wants to sign a contract for a wind project", and he did not let them read the contract. About three hundred people came and signed. He then said, "Bring all of your paper work on your land". There were people there who broke-up that meeting, but he is the *cacique* who is always stealing from this town.

The role and collaboration of the *cacique* is fundamental to facilitating land deals. This collaboration, as hinted at above, has to selectively inform and manage communal assemblies to speed up land deal agreements (Juárez-Hernández and León 2014; Nahmad 2011). I was reminded repeatedly that the *cacique* is the one responsible for the arrival of the wind companies (and the widespread discontent in the town). The *cacique* has an intricate network of friends, family and political allies and is presented as the puppet master behind the curtains. For example, the lead doctor in La Ventos, Dr. Manuel Rios, is the one who all doctors have to report to, and he is also the nephew of the *cacique*. The political arrangement in La Ventosa managed by the *cacique* is widely understood as corrupt. Pajaro, a resident of La Ventosa, explained:

> By now people have become aware that all these politicians just want to get into power. The companies are paying money to the town hall for projects in the community, but many times the mayors will sometimes say, 'we do not get anything from the companies', but the reality is they are pocketing it all for themselves.

This insight into local politics is taken further when discussing political parties in general. I asked Caballo, 'Do you think the *cacique* is making more money here than the COCEI?' They replied: 'It's the same, they negotiate the same. They seem to be fighting, but no. They work together under the table. COCEI, the PRI, they are all together'.

With the *cacique*'s approval, another actor emerges to begin the process of land acquisition for the wind companies: the *coyote*. One testimony became a common description in the town:

> They did not start with a forum. First some people called *coyotes* were brought in, they started talking to some people and reserving the land [for wind companies]. They convinced the *comisariado* [collective land commissioner], they gave him money and then they began to say, "they are going to grow air, because here you cannot grow anything. So you will be harvesting and you will have money'." So they had isolated meetings called by the *cacique* or the *comisariado* and the authorities never called a meeting [to consult the people]. It was isolated. It was done house by house as fast as they could. They signed contracts that they themselves [the companies] drew up. That is how it went.

> **AD:** *What exactly is a* coyote?
> A *coyote* is a person who comes to reserve the land. They are given money to draw up the contracts. So the *coyote* will sell it to the [wind] business people. The *coyote* is the one who makes that deal—the middle man.

A *coyote*'s role is to secure and reserve land for the wind companies at the best possible price, which incentivized higher prices for personal profit instead of benefiting the actual land users. Another land owner seeking more contracts with the wind company to build on their land described the situation with the *coyote* as 'whoever is well prepared will get a good price, but if you are a dumbass, no. He has to negotiate. That is how he gets his share'. According to interviews, land was already concentrated into about thirty percent of the population or less, with roughly 260 land owners in La Ventosa. Nevertheless, the way the wind companies (indirectly) employed *coyotes* to reserve land only further complicated the politics of land acquisition, creating new grey areas in land deals, a kind of 'don't ask, don't tell' in the way land was being acquired. This officially provided the companies a politics of plausible deniability concerning manipulative and aggressive tactics—false promises, changing contracts, preying on language differences, illiteracy, making use of intimidation and other coercive tactics—deployed by the *coyotes* to acquire land (Juárez-Hernández and León 2014; Simon 2013; Hamster 2012; Nahmad 2011). Describing the *coyotes* as 'very subtle',' a woman explains that the *coyote*

> went house by house, and they always worked individually—never collectively. They tried to take advantage the best they could. In fact it was commented in town that people were signing the agreements without the beneficiaries being listed there. So, the owner would sign a contract for 30 years without the beneficiary being specified in the contract or an inheritor in case the owner died.

The *coyotes* created further complications. The same person continues to explain an incident with her cousin who had previously signed a preliminary contract for 20,000 pesos 'to attend some talks and sign a piece of paper'. Later,

> a politician crosses the highway with a piece of paper in his hand and says to her: "sign here, if you want to get 5,000 pesos from Iberdrola", because she had a contract. And she said, "'yeah that is fine'." She signed and the guy said, "'I will be back in an hour to give you money'." [. . .] When she told me this, it was not a year later, the 8th of May and she had not received a peso [. . .], so we went to see Juan Carrasco, representative of the wind company—the boss. Then we told the guy what had happened and he said, "No, it was not 5,000—it was 18,000 [pesos]." She said, "but I have not received a single peso." He said, "well, here is your signature." And her signature was just three letters and the kid [politician] had falsified her signature. So that was theft. That day, a guy was there [. . .], the guy with Iberdrola, currently. When Carrasco came in we told him, we are going to send papers to UCIZONI [a regional resistance group], we threatened him with UCIZONI. [. . .] [W]e forced him to pay her, "we are not going to leave until you pay her," and he did not have all that money, he needed 7,000 more and he sent a kid to his house for the rest. So yes, there were dirty dealings that other workers were making.

The wind companies deny working with *coyotes*, who they call 'political representatives'. While this account likely has other versions, it demonstrates the added complications of local collaborators acting out their personal interest. However, people said in interviews that if one was not a land owner, politician or *coyote*, one was not informed and had little or no idea about the construction of the turbines.

Another technique of land control, lesser known and less discussed than the *coyotes*, was marrying into families. In La Ventosa this idea of 'fatherless children' emerged, which referred to women who married and had kids with wind company employees or locals who returned home or left to work on the next wind energy or other mega projects. This was something a friend told me on my arrival to Oaxaca City, explaining that in Juchitán it was 'common' and that she knew many women in her church that had been left by wind company employees. Apparently, many foreigners migrate to La Ventosa because of the proposed work on wind energy parks and local 'girls' fall in love with or 'are seduced' by the 'tall', 'white' and 'blue-eyed' foreigners. However, an additional narrative built on the 'seductiveness' stereotype of Istmeños women (see Chapter 1), is how these relationships were also arranged by the woman to fulfill a variety of desires: love, travel, companionship and financial security among other personal reasons. With women marrying people associated with the wind energy companies, the companies indirectly gain access to land reserved only for people within the community. Discussing this point, a female civil servant explained:

> Yeah, in general we know it is an economic situation and the companies pay a lot of money, no? So if some foreigner is able to marry and become a part of some family, he has a communal right and they would pass on to become land owners. Yeah, we have a

case like that, we have a case of someone who is not from here and they married a young lady and they have a residence, they bought a lot of land and he is an owner now—he already has the rights.

AD: *Is this common? Is there more than one case?*
Yes there are several cases.

AD: *Could you give me a rough estimation of how many cases there are?*
Let's say five to eight percent.

AD: *Of people in the wind company or people in the town?*
Of the companies.

I was told these marriages left broken homes and created an abundance of single mothers in La Ventosa, when people from the company return to their families back in Spain or move on to their next megaproject. The wind companies created a type of in-and-out migration, which had deeply personal and social consequences for the community of La Ventosa.

The arrival of wind turbines in La Ventosa was orchestrated by the *cacique* and his network of associates—political-business affiliates, *coyotes* and family members—who managed land contracts, wind company negotiations and controlled the distribution of wind energy information. Concern about the *cacique* has existed since before the COCEI, but continues today with the emergence of a counter-political faction in opposition to *cacique* rule, the town hall and the unrealized collective benefits for the town that people were promised and also expected to come from the wind energy projects. This raises questions concerning the resistance to wind energy projects and land deals in La Ventosa.

From Resistance to Negotiations: Land Contracts

When the wind energy projects arrived in La Ventosa, few people had seen them coming and even fewer knew how they would change the town. While resistance to wind energy existed for over ten years in some capacity in the *Istmo*, it was not until the last three years that resistance intensified, primarily in towns around the Laguna facing wind energy development. La Ventosa had opposition before, but was primarily concerned with greater participation in decision-making, benefit sharing and against the overall exploitation of faulty, coercive and manipulative land contracts (Borras et al. 2012; Hall et al. 2015). Nevertheless, effective organized resistance was eventually disabled through intimidation, giving people jobs and paying people money or improving their land contracts. Concerning the latter point, one research participant explained that the land owners will never admit to any negative affectations from the wind turbines: 'they will say "no, no" [damage from the wind projects] because they are receiving quite a bit of money'. However, even among landowners who like the idea of 'clean energy' and their new income security coming from the wind companies, I still heard expressed a feeling that information was withheld from them about wind turbine impact, that they were pressured to sign contracts and that the companies 'refused to negotiate with an organized group', giving the impression of a 'divide and conquer' strategy regarding both private and *ejido* land owners. The land owner, 'Gato Gris', knew that people in other countries were being paid more for having wind turbines on their land and felt they were treated as 'second class people', continuing to say that 'classifications because of our color as human beings should not be a difference for us—the mandate is universal and so should the payment'. The payment inequality from wind companies in the *Istmo* and those in other countries was interpreted as racist by Gato Gris, which can serve as an example of structural racism associated with minimizing project costs and overall overhead.

Money is the primary tool for silencing dissent and dividing people around the issue of wind energy development in La Ventosa. As it was recounted by 'Aguila', a group called La Ventosa Vive 'was practically the entire population', but once the money arrived the groups split.

> [E]veryone started looking for their own interests and that group became several groups and now those groups are divided into more groups and everyone is on their own side, there are some political groups that support, some that do not support the [wind] project and that is why the population is divided. There are people fighting for the projects to continue [construction] in the rest of the *Istmo*, so they can work on it and they do not care even if the [work from the wind] project lasts two or three months, they only want work, but there are others who like the old lifestyle because we used to feel more relaxed without that much crime and life was not that expensive.

This quote gives an idea of the dynamic created by wind companies in the town and how their money became a dividing mechanism. Aguila further explained that groups in favour of the wind energy development in La Ventosa 'think it is progress, even if they have seen on their own population that it is a [work] explosion that goes away quickly, but those are the groups that have seen the benefits and I think they are few'. These groups appear to be enchanted by money, the 'development dream' (Escobar 2012/1995, xlv), alleviating their material poverty and also leading them away from collective concerns of the entire village. The social development funds from the wind company helped to mitigate concerns in the town, reinforcing the narratives of prosperity that will come with wind energy development. This was matched by individual and elite concerns to impose the wind projects at any cost on the town, which includes tactics of intimidation and sometimes violence. '[W]hen somebody does not want the wind energy projects they [the *cacique*] intimidate them using people with firearms', explained Dona. While Aguila said they 'know two or three people who have been intimidated, people have come to them and said: "Either you let the project go ahead or there will be consequences for you"'. People were hesitant and at times afraid to talk about the details of wind energy in the town, and that there are gunmen related to local families who are in every town and are working with the unions currently fighting for more wind parks.

While violent intimidation was present, money proved a more effective weapon in the context of La Ventosa. In one notable case, a land owner who was opposed to the wind energy projects and a member of the Association of Indigenous Communities in the Northern Zone of the Isthmus (UCIZONI)[2] was approached twice by the wind companies with offers to lease his land. He rejected them twice, identifying as a farmer who took pride in his work and was far from enchanted with the new changes to La Ventosa. Nevertheless, he eventually signed a contract with the companies. When I asked the reasons why, he explained:

> Two things. What mostly made me sign the contract is my land is in the middle of all of the neighbors—they left me all by myself. Two companies came and talked to me and I did not sign anything, I was the only one who did not sign anything. Why? Because I did not want to. Like I told you, I did not want to lose my land, but I was all by myself and they said, '"Are you in or not?"' And they do not care if I am in or out [of the negotiations], so I had to give in. If there was a way for me to do something, to farm and for that to be productive—I would prefer that. In fact, if I had not accepted once the work started in that area I would not have been able to stop it. So the things the companies do are going to affect me. Why? Because the road belongs to everybody, they have the same right that I do. It is not in my interest to end up all alone. That is why I gave in. For my family and for this reason.
>
> **AD:** *So essentially, you sold your land because you were going to be surrounded anyway and in the end it is safer?*
> More or less. I am not so convinced [that it is safer this way]. For me not so much, but everyone wants to have a wind turbine on their property—not me—I had to play the game. I prefer ox carts to cars—that means I prefer to live in a time of ox carts.

Another female *ejidatario* felt similarly, saying that 'if 70 percent said out with the wind energy I would be with them, but if we are only 30 or 29 [percent] then no'. This instance shows how these industrial projects control and force people to ultimately cooperate with the wind companies and their surrogates even when they vehemently oppose such project developments. The following chapter will expose a similar situation in the Bíi Hioxo wind park, where another land user refused to surrender his land, lifestyle and a wind turbine development.

In the last few years, resistance in La Ventosa was reignited with a counter-town hall to resist the power and corruption of the *cacique*. This counter-political faction was able to prove with photos, video and political recipes that the *cacique* and his people were engaged in voting fraud. This distribution of funds for voting credits, it was explained, was made possible by the money distributed to the town by the wind companies. They made an appeal in the courts in Oaxaca City and then in the courts in Xalapa with little result. Nevertheless, the townsfolk persisted in blockading the highway, occupying the city hall and leading the protests until the Mayor of Juchitán, Saúl Vicente, was recalled saying: 'You can continue going to war or I can offer you this tie so there can be peace', offering to divide political control of the town into two factions—the *cacique* and counter-*cacique*. This 'practice of "shared" municipal administration,' Rubin (1997, 52) was a political technique widely deployed by General Charis and 'was one of the ways in which Juchitecos negotiated the presence of the PRI' in the *Istmo*. For about two years La Ventosa has had a shared municipal administration. On my first visit to La Ventosa, a local political candidate explained that the *cacique* 'is the one who rules the wind energy project and the people who raise their voices are intimidated by him, but we are not intimidated because we have courage to denounce it'. This faction is largely concerned with the failure of social development and benefit sharing in the town. While many people in the town criticized them for going after the wind energy money allotted to the town hall, they also provided me with an important opening for this research and criticism of the wind energy projects. Self-censorship was inherent to the interviews ('people are afraid to talk about the wind turbines because of the *cacique*'), but this faction created a space for discussion and made it possible to conduct interviews with people in both political factions. Said simply, there was another gang in town that opened space for conversation, criticism and action against the wind companies and cacique. After covering the arrival of the wind turbines, the local politics and land deals, the focus now turns to the environmental and social impact experienced by the residents of La Ventosa.

ACCOUNTS OF ENVIRONMENTAL IMPACT

Ecological impacts were frequent occurrences in interviews. The extent of this impact varies depending on each project location, but there are a general set of environmental damages to be expected. Animal habitat loss is the first as wind turbines, especially early generations, cannot have trees interfering with the wind or blades. Road construction to access the wind turbines also necessitates the clearing of tress, bushes and vegetation in addition to soil compaction. This results in the immediate alteration and destruction of habitat in the area of wind parks. Second, road construction is then compounded by the construction of wind turbine foundations. Ground water in the coastal *Istmo* is anywhere from one to three metres below ground level, while wind turbine foundations are generally between seven to fourteen metres (thirty-two to forty-five ft.) deep and about sixteen to twenty metres (fifty-two to sixty-eight ft.) in diameter, thus requiring that the ground water is transformed into concrete. These foundations, combined with the road construction, resulted in extreme flooding during the wet seasons (April–September) (Dyer 2009). One landowner who has a contract with the wind companies explained how the road dams up water for '15–20 days and that washes into my land and that prevents me from working it'. Research participants report that flooding in the wet season corresponds with extreme drought conditions during the dry seasons (October–March) as a result of aboveground land alterations changing the soil composition, and underground concrete and electrical infrastructure disrupting the water table drainage, as well as the constant spinning of wind turbines. These environmental changes have complicated farming in the area. People told stories of wells drying up and commented that the wind turbines have altered local weather patterns. The latter claim surfaced in other research in the area (Simon 2013) and is a negative environmental impact supported by wind turbine environmental impact studies (Tabassum-Abbasi et al. 2014; IPCC 2011; Wang and Prinn 2010; Keith et al. 2004). A teenager I met on the street exclaimed: 'The wind turbines have scared away the rain and that is why the grass and animals are dying', which raises more questions than it answers, and begs for more research from climatologists and natural ecologists.

The *Istmo* is located along one of the principal bird migration corridors in the world (Juárez-Hernández and León 2014). This has led to reports of extensive bird deaths in the area.[3] It is already well known that birds lose control when in range of wind turbines, which results in high death rates (Tabassum-Abbasi et al. 2014; Drewitt and Langston 2006; Barrios and Rodriguez 2004). This is compounded by the clearing of habitat for wind turbines, roads and their monolithic foundations. A World Bank sponsored study conducted in the La Venta II wind park between 2007–08 discovered migratory bird mortality rate of twenty or more birds per installed wind park megawatt per year (Ledec et al. 2011)—a figure that is said to exceed the average wind turbine-bird death rates. Wind companies have devised some mitigation methods to reduce bird mortality rates: increasing the visibility of blades with ultraviolet paint and deflectors; installing transmission wires underground; timing of construction and maintenance; siting wind turbines as close together as possible to minimize their presence; and stopping turbines during bird migration seasons (Tabassum-Abbasi et al. 2014). While some of this is designed to be integral to future wind park projects (IDB 2011), there is still a looming concern over bird deaths as obtaining accurate measurements is a challenge, due to access to the park sites, scavenger removal and the sheer size of wind parks (Tabassum-Abbasi et al. 2014). Furthermore, I was told by a research participant that 'the people who work maintenance [on the wind parks] have to pick them up [dead birds] and sometimes they bury them so nobody can see them'. When I asked what motivated the companies to behave like this, the person replied: 'Because when they began they said they would take care of the animals and the project was not going to cause harm to neither people nor animals, but that is a lie because we have found several dead birds under the wind turbines'.

These large-scale wind turbines also leak oil. Wind turbines use oil to lubricate the propeller gear boxes. This oil leaks into the ground and subsequently into the shallow ground aquifers and water wells. One woman asserts that 'the wind turbines are contaminating the water and you can now see oil in the wells,' while another stated that '[t]he soil has been contaminated with gasoline and oil'. I had also seen this outside Juchitán in the more recently-built Bíi Hioxo wind park discussed in Chapter 3. It was brought up in one interview that the wind turbines give off a 'strong' and 'ugly smell that comes from the oil'. This raises the question of the wind turbines' proximity to people's homes and the fields where they live and work. The most revealing interview concerning the leaking oil was with a proud and enthusiastic maintenance worker wearing wind company paraphernalia. I met him when I was conducting door-to-door interviews. When I asked him what kind of work he did, he replied:

> Excellent [work]. They are good public works that are a benefit for the town. The only thing they say is that the wind turbines are bringing illness because they leak a lot of oil. You know how people talk. They say that oil causes cancer that is the concern people have.
>
> **AD:** *You who have been working on these projects and know them, do they have an environmental impact? What have you seen?*
> I do not think so.
>
> **AD:** *What part of the wind turbines do you work on?*
> Actually in my case it is the construction project.
>
> **AD:** *Oh, so you are working on the foundations?*
> Yes, the base of the wind turbines.
>
> **AD:** *So you have seen the oil dripping?*
> Yes, right now. Now you can see it. Several of the turbines you can see have oil leaks. If you want to go out in the sun you could see several—thirty or forty percent are leaking oil.
>
> **AD:** *So besides the oil, do you think there is any other way the turbines could be hurting people?*
> I do not think so. Because well, up to now—nothing. Honestly there have been some accidents, which is normal in any construction project.
>
> **AD:** *We have heard about wind turbines' blades falling off.*
> Yes, some. Especially over in La Venta. There was one, well, you know, they burn because of lightning.

This man's blunt honesty was surprising and insightful. His account not only confirms oil leakage, but also demonstrates people's perceptions concerning oil contamination.

Wind turbine fires are another environmental concern. I heard of unverified accounts of three or four wind turbines burning, and saw the remains of two turbines in the surrounding La Ventosa area. This was significant, as not only did it provide accounts of fear from the wind turbines with people describing the way they sway in the wind, but they also made people 'sick'. One exploded right next to the town to the north: one person became upset when he told me about how his nephew got sick with a fever and was vomiting, attributing this to the smoke from the turbine fire. These explosions were the result of either lightning or a strong south wind that overwhelmed the wind turbine.

There are also animal deaths caused by wind park development and construction. In general, the children expressed the greatest concern about the loss of the animals and habitat destruction near their school. The loss of bushes, brambles and trees for road and wind turbine construction have, I am told, resulted in a loss of iguanas, snakes, badgers and rabbits. Asking a young child how he thinks the wind turbines kill animals, he replied: 'The electric current grabs them', giving me an example of how 'rabbits dig holes in the ground and sleep in a hole and the electrical current comes to them and burns them', asserting that he has seen many dead rabbits since the arrival of the wind turbines. While I had difficulty understanding, a retired engineer explained this could be possible from poor electrical grounding of the wind turbines and the residential homes in the immediate area. That could make sense, considering the possibility of relaxed regulations in the area. Nevertheless, these numerous accounts of animal deaths where not restricted to La Ventosa alone, which remains an open question for further research and investigation.

More significant were cow and cattle mortality. The general explanation is that oil and other fluids (such as blade waterproofing resins) leak from the turbines and blades into the soil and water and the bovine animals graze and eat from these toxic grasses, which includes drinking contaminated water. It was common to hear comments such as: 'Sometimes I hear that the cows lose their young and that is a problem. Then they cannot get pregnant—that is what I heard'. Another person explained:

> When a cow or an animal dies from an unknown cause, they dig a hole in the ground and they do not eat the meat and they bury the animal. My grandfather is 86 years old and said, before not as many animals died from unknown causes as they do now. They [people in the village] attribute this to the increases in wind turbines and my grandfather thinks that way, but I haven't seen it.

A similar story emerged from a pastor[4] who was gifted a calf. He had friends who were watching the calf among their other four cows, who all suddenly died without explanation. He explained: 'They said that has never happened to them before. . . . All of a sudden they died and they do not know why and the veterinarian did not give a diagnosis as to why. They just died'. Eventually, while we were walking around the market in an attempt to talk with and interview people, a woman walked up and when asked if she was aware of any illness coming from the wind turbines she replied: 'two nearby ranches' fertility have been diminishing in the cows. We can prove that because there were years when there were no calves born. In my ranch it has been happening also'. While she said other people might have other experiences, '[t]he cows pregnancy period is nine months and when that time is up there should be newborns, but I have noticed that since August until now, no cows have gotten pregnant'. Notably, this woman was critical of the wind energy projects, but did not reject their existence, as is common in other towns around the Laguna. Instead, she argued that there was no 'true compensation for the environmental effects' of wind turbine', implying that with social and ecological compensation the wind energy projects could benefit the town. Fourteen interviews mentioned cow deaths, illnesses and infertility, but more importantly it appeared as common knowledge among cattle ranchers that wind turbines were responsible for all these serious problems.

The perception of people concerning wind energy environmental impact is alarming and resonates with the literature (Havas and Colling 2011; Bakker et al. 2012; Farboud et al. 2013; Jeffery et al. 2014; Tabassum-Abbasi et al 2014; Evans 2014). While such accounts from above people are limited, they still present serious concerns whether directly or indirectly related to the wind energy projects and their damaging infrastructure. Additionally, wind turbine impacts increase when they are located next to water and sea life, which was not the case in La Ventosa, but remains a principal concern with the wind parks built and being planned outside Juchitán (Chapter 3) and the towns around the Laguna Superior and Inferior (Chapter 4).

LAND CHANGE: INEQUALITY, RURAL GENTRIFICATION AND OUT-MIGRATION

The majority of responses from people during interviews can be best summarized as follows: 'There are no real benefits' for the town, and only the landowners benefit from the wind projects. Out of sixty-three interviewees, forty-seven said there were no social benefits while thirty-six said only the landowners benefited. 'Social benefits' refers to collective benefits achieved for the people as a whole—a community. It was not only the landowners who benefited, but also political authorities such as the *cacique* and his network of associates. There were twelve people who felt they were not affected by the wind turbines, viewing these projects as generally beneficial to the town even if these benefits were admittedly limited. The best explanation for selling land came from Elder Gato, who said: 'It is hard to steal a wind turbine, but you can steal a cow'. Elder Gato was among at least two other landowners who felt grateful for the wind companies, even if at moments during the interviews he expressed anger at the wind company negotiation practices, profit sharing (compared with other countries), and a lack of information about the ecological impact of wind turbines.[5]

These land owners felt the wind companies provided new opportunities for the town, themselves and their families, allowing them to send their children to university. The same narrative is articulated by Howe and Boyer's (2015, 35–36) 'Don Julio': however, such expressions of appreciation were immersed in discontent and at times articulations of utter powerlessness—'they [the wind turbines] are already here, what are we going to do?' Walking around the town, one can see from people's homes and automobiles—American brand name SUVs and pickup trucks—who works with the wind companies, and who does not. Benefits from wind energy projects included: (temporary) work, (some) paved streets, a market centre, the house of culture, the painting of schools and the building of a soccer field. One mother told me that the wind companies helped low-income children with painting lessons, Zapotec language classes, and summer school—not to forget classes about wind energy in primary schools as well as technical courses at the University of the *Istmo* to train people in electrical engineering. However, many of these civil projects, I was told by the counter-cacique faction, were achieved only through continuous struggle and protest.

While benefits were real for landowners and political leaders, discontent with the wind energy projects seemed overwhelming overall. For example, regarding improvements to the schools:

> When Iberdrola was about to enter a few years ago they said there would be social benefits. This is what the companies said, they were socially responsible, and it was even their slogan. So the first thing they said was that they were going to improve the school because the school is 40 years old and it expired 15 years ago and until now there is no progress. The only thing they did was brought two paint cans to the school and took a picture and they said, "We are supporting education." They brought two footballs; took another picture and said, "We are supporting sports." So there is no benefit for this town—there is none.

Accounts like these were common. While some appreciated the four to twelve blocks of paved roads in the town, other felt this construction was an utter joke—'[S]o the streets were paved, but the only people who benefited are those with land'. Frustrated, Dona points out that the running water 'is not drinkable and they put some pavement on the streets, but they did not take care of the sewers and the water, they just did things to make the town look pretty. At a certain point this hurts us'. Considering the scale of the project, the money involved and people's quality of life, they felt it was a change for the worse. At the end of an interview a woman summarized: 'We are still poor and now we are surrounded by wind turbines'.

Temporary Work

The employment provided by the wind energy companies was met with similar attitudes. Those seeking wind park jobs were in competition with migrant laborers and other specialists who came from overseas to work on the wind projects. Work was limited in duration, quantity and went to more highly-skilled foreigners due to their technical expertise (Simon 2013). Time and again, people stressed: 'They promised work, but nothing—it is worse than before'. The jobs available for locals were temporary: anywhere from three months to a year and a half. Gaining employment also depended on people's relationship with the *cacique* and his union networks. Repeatedly, I was told that jobs became a way of buying votes and silencing people. Likewise the idea is that '[i]f you want to keep your job, you have to do everything the company says'. People are even paid, according to interviews, 200 pesos to agitate and intimidate people critical of wind energy at the Free, Prior and Informed Consent (FPIC) consultation that began in Juchitán in November 2014.[6]

Electricity

Dissatisfaction with work was combined with rapidly rising electricity rates. Because of land use changes for wind turbine construction, La Ventosa is now considered an industrial town, meaning it loses all state subsidies and residents are forced to pay industrial electricity rates regardless of the fact that they live within the power plant itself. One man asked: 'How is it possible that they consider us an industrial zone because our region is producing energy they are taking to other countries?' Thirty-seven of sixty-three people interviewed were infuriated by rising electricity bills. 'I feel that we should not have to pay electricity bills in this town because we are surrounded by these wind turbines', says a mother. While another person explained:

> Every two months the electric bill goes higher. So when the bill comes in at 800 or 1,000 pesos the farmer does not have enough to pay for that bill. So then the CFE comes and cuts off your power because you have not paid. Poor people. If it was generating electricity our families should be doing well, we should be able to enjoy that, but there is no benefit.

A small sandwich shop owner said they pay 3,000 pesos (approx. USD 146) every two months, which represents a huge jump in electricity bills that risk putting them out of business. Another person contends they are paying 1,000 to 1,200 pesos every two months for electricity in their home, a pressure that is justified by the change in soil use for wind turbines. Even during an interview, I watched in the background as a CFE employee got out of a truck, knocked on the neighbour's door and afterwards walked around the property—I assume they were shutting off the electricity. Then, during another interview just down the street, thirty-five minutes later, the CFE was still driving around and going house-to-house and a woman yelled: 'We want light! We want light!' The situation appeared dispiriting and disheartening. Enclosed by wind turbines, the electricity prices for residents were skyrocketing, with residents telling me that ninety percent of the jobs made available by wind companies had since left the town. La Ventosa, like other rural towns in Mexico, already had difficulties with political corruption, poverty and income generation, but the arrival of wind energy development began instigating a type of rural gentrification in the town.

Rural Gentrification

Two decades or more ago, Martin Philips (1993, 138) argued that 'rural studies would appear to lag behind urban studies in recognizing the diversity of ways one can interpret and understand gentrification'. More recently, Darren Smith (2011, 599) affirmed this, stressing the increasing resonance of gentrification in rural studies. Arising from public regulation and private investment, gentrification is a dynamic process of urban revaluation that creates price hikes on space, and by extension, property, in targeted areas (Lees et al. 2008). Smith (2002, 390) makes a call to 'widen the spatial lens' of gentrification studies, while Davidson and Lees (2005,1170) define four foundational characteristics of gentrification: (1) reinvestment of capital; (2) social upgrading with in-migration of high-income groups; (3) landscape changes; and (4) direct or indirect displacements. While characteristics of rural gentrification have already been demonstrated with large-scale capital infrastructure investments, exclusionary land leasing practices, in-migration of high-income groups (employees)[7] along with steep rises in electricity prices, wind turbine development has given rise to a new geography of rural gentrification—for example, its role in driving up land and rent prices. As a civil servant explained:

> When the companies first came they arranged a rental contract and they came to look at the lands and they put in a clause that when the owners were ready to sell, they would have to sell to the companies. Then all of the prices went up more than 200 percent, the price of land then was 4,000, 5000 pesos. Now it costs more than 50,000, 60,000 pesos [per square hectare].

Talking about the change in rent for an average dwelling, an *ejidatario* says, '[i]f the rent before was 300 pesos, it is now 3,000 or 4,000 pesos—that is too much'. This influx of people coming to plan and execute the wind projects, which was significant for the size of La Ventosa, had varying effects. The arrival of more people and money triggered a moto-taxi rush, with around 200 people buying moto-taxis, which now sit in people's houses after wind park construction ended. 'The companies paid more than what the people were used to here, but it had serious consequences once the wind energy companies departed. All the moto-taxi drivers wanted to charge the same amount [of money] to everyone', explained one driver. This trend subtly influenced restaurants to accommodate foreigners and to take advantage of the new money in town. Discussing cultural changes, a human rights defender in Tehuantepec pointed out: 'If you go to La Ventosa, there is a restaurant there and it used to be you could eat garnachas there, cocada, torta, coffee, but not anymore, now it is all gringo food. Light skinned people like you, more or less, go there and they have their menu there'. While this change might be welcomed by some, to others it signifies a subtle colonial erasure. This is compounded by changes in the quality of food. Emerging from a discussion about the 'sickness' in the town one woman explained:

> We think it is something coming from the food. Actually, before we did not eat this kind of chicken, we ate chickens that were raised. The pigs are the ones now that just eat [industrial] feed and before that was not the case. Before we just ate animals that were roaming around, eating whatever they could find, like corn [stalks]. Now they are just purely eating feed made of chemicals and people are getting sick—a lot.

The shift from farming to 'wind harvesting' has had significant ripple effects, forcing people to import their food, slowly integrating the population into food logistics and transportation which comes with new higher overall costs (see figure 2.2). 'Every day the price of meat gets higher', a woman complained, telling me that it is over 140 pesos (approx. USD 6.5) per kilo. This impact was multiplied by other significant cultural changes with the arrival of new people, habits and out-migration in search of work.

Figure 2.2. Mural in Juchitán: 'Autonomy and freedom of expression—My totopo will not have genetic modifications—Cacique oppress you—Political parties divide you—Enterprises invade us.' SOURCE: AUTHOR

Crime and Drug Consumption

'[S]ince the arrival' of wind energy projects, said a young woman, 'the problems that were already here started to grow with increases in drugs, in rent and food [prices]'. Crime and drugs grew with the arrival of wind energy parks. Explaining the rise in crime in the town, people continually blamed it on 'outsiders' or 'foreigners' who think La Ventosa is rich because of the wind parks. 'Because there are a lot of people who have wind turbines on their land and they have higher earnings, there have been more robberies, assaults and people have broken into houses', explained the woman, continuing: 'So there are a lot of people coming from the outside, so since this town is so small we all knew each other, but not anymore'. Twenty-two of sixty-three people recognised a rise in crime while thirty felt there was a visible rise in illicit drug consumption. A civil servant explained how this became a serious issue for the town:

> It was about three years ago, when the high school, the teachers, the workers and the staff of about 30 workers and three-hundred students demonstrated and went on strike and closed down the road. Because of drug consumption, the population was living in a situation of insecurity. Young people from outside the community, but from the region, were coming and selling drugs, selling coke [cocaine]. So we demonstrated before the

government so [. . .] they would give us more security. And that these young people that would come and distribute drugs would retire and go away because they were practically poisoning our students.

I followed up by asking if this rise in drug use had 'some relationship with the wind companies?' The civil servant replied: 'Probably, yes. Because before there was delinquency, yes, but not so much like when the foreigners arrived'. Stories kept emerging, specifically in interviews with mothers, about how they watched young people intermingle with wind company workers, who provided a gateway into their drug use. Summarizing the situation, Aguila said: 'The jobs leave, but the drugs stay'. The FDI and wind energy development created not only a direct rise in housing, electricity and food prices, but might support a micro version of Paley's (2014) *Drug War Capitalism* theory—where economic growth correlates with drug consumption and production. Meanwhile, drug cartel related violence increased in the *Istmo* against the backdrop of wind energy development and resistance.[8]

Out-migration

Wind energy development triggered a duel process of revitalization for some, and ghettoization for the majority, intensifying pre-existing negative relationships and patterns within the town. This gets expressed in increased out-migration. The dramatic rise in prices, as well as the psychological and physical discomfort from living surrounded by wind parks, has intensified a poverty trap, resulting in indirect displacement related to land change. Summarizing the situation, a pastor explained:

> People complain because they cannot work the land like they use to, even though they were given money for the renting of the space, but they cannot work in agriculture. There is no more corn, beans, or watermelon. The ranchers are trying to make the best out of the little bit of grazing land they have, but there is no more production in agriculture.

This recent change from agriculture and livestock to wind energy dependency appears to be slowly establishing further dependency on the importation of food from industrial sources, reflecting larger macro-economic trends in Mexico and insecurities associated with dissolving small-holder agriculture (Bello 2009; Schutter 2011; White et al. 2012). This has widened the income-inequality gap in La Ventosa, allowing a kind of rural gentrification to flourish and encouraging existing out-migration for work. A middle-aged man elucidated:

> Yeah, much less work—less work. That is when people will start to migrate as far as the United States, where people normally migrate to or maybe to other states or simply they will stay with the possibility of living day by day. For this reason it hurt their children in the educational field: they will not be able to continue their study because of the lack of resources.

AD: *Do you see this starting to happen now?*
Yes. Yes, I have a lot of friends who do not study anymore because their parents have ended up without work and they have migrated to other states and started working and plumbing in bathrooms and as low-paid workers in tourist zones where there is a little more work.

While these trends—land change, rural gentrification and out-migration—have already occurred, it is thought they will worsen in the short-term future. The extent of out-migration requires further investigation, but what is for sure is that the land owners and elites benefit, while others feel they 'are still poor and now we are surrounded by wind turbines'. Wind energy development results in a series of dependencies, notably on the construction of more wind energy projects for the landless workers (see Borras et al. 2012), pitting them against their neighbouring coastal communities who not only survive from farming, fishing and selective engagements with the local economy, but have taken up an insurrectionary stance against wind energy development on the Laguna. Wind energy projects have far-reaching social and environmental impacts which public relations, market opportunities and the hope that they will mitigate ecological catastrophe appear unable to dispel.

CONCLUSION: GAINS, LOSSES AND STRANGULATION

There are significant gains for the *cacique*, the political authorities and landowners. There are token gestures of social development within schools and also local infrastructural development. However, there is far greater discontent and damage to the people and environment of La Ventosa. Steeped in systemic political corruption, land grabbing is executed by local middlemen backed by wind companies to acquire land through leasing contracts. A good portion of the town, even those sympathetic and willing to work wind energy are rendered discontent, jaded and apathetic—'even if we are bothered sometimes, people cannot do anything and you know how politics works [. . .], the population does not decide what is going to be done and that is why we *conform to the things we are given*'.

The concentration in and enclosure of La Ventosa with wind parks is either directly resulting in symptoms associated with wind turbine syndrome—noise irritation, exhaustion, insomnia, headaches, dizziness, and so on—or indirectly contributing to a variety of other health issues that remain undetermined and require further research. This shift takes place in a context of large increases of investment and cash circulating in La Ventosa which has taken the collective form of rural gentrification that has had a variety of social and cultural effects, and has intensified and made worse pre-existing issues in the town, such as poverty, crime, drugs and high unemployment. This accumulation has resulted in an increased out-migration from La Ventosa to find work in other states in Mexico and the United States.

While these projects have no doubt benefited some, they have come at a great collective and individual cost, thus intensifying the daily struggle. Reflecting on change in La Ventosa, a woman explains:

> [i]t used to be beautiful, but not anymore. Maybe you have seen on television somewhere that La Ventosa is a famous place because of the wind turbines, but those things are harming the town. I live here and I cannot stop thinking about how much the town has changed, it is not the La Ventosa that I grew up in. Now there are a lot of cars and people entering the town from every side, we are not safe anymore. Here in La Ventosa the kids use to play in the dirt, every one of us did it, but now that has changed a lot, you have to take care of the kids—you have to watch over them.

Similarly, every now and again one will meet an irate elder who contends that life 'was more beautiful!' without lights. Wind energy business emerging from climate change concerns and mitigation policy is now negatively altering the new areas sited for wind energy development. The clean development mechanism (CDM) among other market mitigation schemes is subsidizing transnational corporations such as Walmart, Bimbo, and Amstel to continually expand their business, furthering the spread of plastic trinkets, processed food and beer across the world in the name of slowing anthropogenic climate change. Wind energy, while less damaging than fossil fuel, nuclear and hydraulic fracturing and other natural resource extraction models, has now expanded industrial development and energy projection in a manner that impacts clearly, and negatively, on people who live within and around these wind power plants. Softer, friendly and 'green', wind energy has created a situation where people can live within wind power stations. The greatest accomplishment of industrial-scale renewable energy production has not been to alter the trajectory of industrial development and its consequential environmental degradation, but to expand it into the lives of people at the periphery of the national and international economy. The next chapter looks at the expansion of wind energy development in communal lands outside Juchitán on the Laguna Superior, where people have organized to resist the Bíi Hioxo Wind Park.

NOTES

1. The Renewable Energy and Energetic Transition Law (2008), The Special Climate Change Program 2009–2012 and The General Law on Climate Change (2012) that seeks to reduce emissions by thirty percent in 2020 and fifty percent by 2050 based on year 2000 emission levels. This plan is known as the 10-20-40 vision.
2. *Unión de Comunidades Indígenas de Zona Norte del Istmo.*
3. See also Simon 2013.
4. This pastor I was told had a vested interest not to be critical of the local wind companies.

5. This included a feeling of discrimination and racism arising from negotiation practices, attitudes surrounding Zapotec life and unequal benefit sharing. James Anaya's (2015) observation of the FPIC procedures regarding the next wind park projects also commented on this.

6. See Friede and Lehmann 2016.

7. Most people migrating from Northern or Developed countries with wind company jobs will resemble a high-income group.

8. This raises further questions in this region regarding Paley's (2014) *Theory of Drug War Capitalism* documenting how cartel violence terrorizes rebellious areas, which is used as a justification for the military to invade and further terrorize activists fighting development projects.

3

Counterinsurgency for Wind Energy: The Bíi Hioxo Wind Park

On 24 February 2013 in the Seventh Section neighborhood of Juchitán de Zaragoza, the Popular Assembly of the Juchiteco People (APPJ)[1] was formed. The next day, on the southwest outskirts of town, they went public with a barricade on the highway to Playa Vicente to halt the construction of the Fuerza y Energía Bíi Hioxo Wind Park. On 26 March, 1,200 police arrived to break the barricade (Earthfirst! 2013). A member of the APPJ recounts the event:

> When we left the barricade heading north, about halfway we saw between 26 or 30 buses of state police, then my sister said to me: 'Let's go back, let's go back!' And I told her, 'No, we have to keep going—stay calm.' She continued: 'They are going to arrest you,' and I answered, 'No, keep walking.' Then we got in my truck and we drove past the police convoy, arriving to a place called El Tanque. When we arrived in town, most of the neighbors were standing on the road—there were a lot. People were all over the road asking: 'What can we do? What can we do?' [A] neighbour close to me asked what to do, I told him: 'Grab rocks, grab sticks, and anything you can to block the road, because they already passed, but they are going to need this road to get out and then they will be fucked.' To the other *compañarxs* who were wondering about what to do, I told them to go to the loud speakers in the town to make the announcement, and I went to do the same. I went downtown where the speaker was on a platform, so I called from there and said: *'Now is the time! We have to defend our land; we have to defend our territory. My fellow countrymen, Women, kids, elders and men grab everything you can: sticks, rocks, machetes and anything to defend our barricade!!'* So, the mayhem starts. People came from the Fifth Section and downtown, they were young people, women, and children. Some women, even those who were identified for belonging to the [local leftist political party] COCEI tied their *chals*[2] on their waist as if it was a *ceñidor*[3] so they could fill it with rocks and with that they were running to the highway.
>
> The police shot tear gas and slingshots because some of them are from Juchitán. And the people were defending themselves with anything they had: stones, throwing sticks, etc. The police tried to recover their vehicles, they couldn't. The people saw them coming, burned a truck, and started fighting them. There was barbaric fighting. When the police tried to take back their vehicles from the barricades and run away, the people

attacked them with stones and made them get out of the vehicles and run into the field. They ran towards Unión Hidalgo and left their shields—we picked up around 40 shields that the police abandoned. At night, somebody told us that the police were picked up around Union Hidalgo by the military because some of them were naked in order not to be identified as police.

This battle went from twelve in the afternoon until two in the morning. Injured and wounded people avoided taking ambulances to avoid being taken into police custody; the resistance responded to the police arrests by capturing a police woman to negotiate a prisoner exchange with the help of a local priest. Further, the APPJ negotiated 130,000 pesos (approx. USD 6,352) in exchange for geographical survey equipment seized at the barricade for medical bills incurred during the battle, later leading to accusation of extortion by the owners—a charge dropped after investigation (PBI 2013). The *Binnizá* (Zapotec) and *Ikoot* (Huave) farmers and fishermen associated with the APPJ, the Assembly of the Indigenous Peoples of the Tehuantepec Isthmus in Defense of Land and Territory (APIITDTT), and others, have asserted their perceived and often disputed customary rights that have proven a threat to wind energy projects in the region.

The account above offers only a small glimpse into the level of conflict and resistance generated by wind energy projects in the Isthmus of Tehuantepec region (*Istmo*) of Oaxaca Mexico. This particular case fits a wider, well-documented pattern of so called 'green grabbing' (Vidal 2008; Fairhead et al. 2012) that describes how land is grabbed and controlled in the name of an environmental ethics to promote sustainable development and climate change mitigation programmes (Peluso and Lund 2011), consequently reigniting old and creating new conflicts over the people's land and natural resources. As argued by Dunlap and Fairhead (2014), there is a history of, and continued reliance on, military techniques of counterinsurgency to advance natural resource control and industrial development in ecologically diverse and ecologically sensitive areas (Peluso and Vandergeest 2011). This reliance on counterinsurgency and other militarized techniques has been well documented in relation to nature conservation as captured in the notion of 'green militarization', defined by Lunstrum (2014, 817) as 'the use of military and paramilitary (military-like) actors, techniques, technologies, and partnership in the pursuit of conservation'. Büscher and Ramutsindela (2016, 3) extend green militarization to a wider notion of 'green violence', describing the material, social and discursive dimensions as well as the longer histories of violence and militarization in a particular region and its conservation areas.

Green militarization and violence is not limited to conservation parks, but can also be found with any project utilizing an environmental rationale and repressive techniques to impose its existence on an entire (or segments of) population and the lands they inhabit. This chapter builds upon the notion of the *greening of counterinsurgency* outlined by Dunlap and Fairhead (2014), which roots 'green violence' within a specific military doctrine and history emerging from (asymmetric) colonial wars that has become increasingly popular, gaining widespread application within militaries (Owens 2015), police departments (Williams 2007/2004; Williams et al. 2013), resource extraction companies (Rosenau et al. 2009; Jarvers 2011; Gedicks 2015) and even marketing agencies (Copulsky 2011). Not only does counterinsurgency establish a strong historical link with previous military, paramilitary and police operations within a region, but it also unravels the 'conflict management' approaches of dominant public, private and non-governmental actors (Verweijen and Marijnen 2018; Duffy 2016; Dunlap and Fairhead 2014; Peluso and Vandergeest 2011). The following case of Bíi Hioxo Wind Park illuminates the relevance of adopting counterinsurgency as a lens to analyze efforts to break popular opposition to wind energy development in the name of mitigating anthropogenic climate change.

Planning for the Bíi Hioxo wind park began in 2006, it was among the first wind parks planned, and eventually completed, on the Laguna Superior, in October 2014. This wind park became the third largest in Latin America with 117 wind turbines and a capacity of 234 megawatts (MW) (CDM 2012; GNF 2013a). Registered with the Clean Development Mechanism (CDM 2012, 2), which claimed that the Bíi Hioxo wind park 'will make a positive contribution to the protection of and care for the environment, directly addressing problems of climate change,' and 'does not require extraction, drilling, transport of fuel or water consumption, and does not generate polluting dangerous or radioactive waste'. This project is spread over 2,050 hectares of communal land (*comunales*), had USD 433 million invested into the project, and participates in the 'self-supply' (autoabastecimiento) wind parks regime in the region (Bnamericas 2015; CDM 2012). Self-supply electricity is private, generated and reserved for the investors or co-owners of the project, facilitating the commodification and privatization of wind energy resources. The leading investors and project developers for this project are Gas Natural Fenosa, Cementos-Moctezoma[4] construction, Tiendas Chedrahui superstores, and Saint-Gobain México Construction among other companies (CODIGODH 2014; CDM 2012). Fuerza y Energía Bíi Hioxo S.A. is a subsidiary of Gas Natural Fenosa (GNF), a member of UN Global Compact[5] and valued by 'the main sustainability indices' and maintains 'its firm commitment to respect for human rights and specifically the traditional ways of life' (GNF 2014, 6, 224). Considering the level and intensity of resistance described above to defend a barricade attempting to halt the Bíi Hioxo wind park, these claims appear questionable, if not disingenuous (GNF 2013a).

Based on participant observation and twenty recorded informal and semi-structured interviews focused on the Bíi Hioxo wind park, this chapter begins by providing some background on counterinsurgency warfare in rural, and often Indigenous territory, touching on the 'property based approach to counterinsurgency' and its history in Oaxaca (Bryan and Wood 2015, 149). The following section looks at the arrival of the Bíi Hioxo wind park in the area, later examining the 'hard' and 'soft' counterinsurgency techniques used to undermine resistance and complete the wind park on communal land. The next section reviews the reported environmental impacts experienced by inhabitants living and farming within the wind park after one year of operations. The chapter concludes with the assertion that the Bíi Hioxo wind park is causing violent social divisions amongst people, altering the subsistence and cultural practices of Indigenous groups seeking to uphold traditional ecological relationships. Here the concept of 'cultural genocide' is introduced to capture the seriousness of the processes facing people seeking to maintain a relationship with their land and culture.

COUNTERINSURGENCY, SOCIAL PROPERTY AND INDIGENOUS PEOPLE IN SOUTHERN MEXICO

Counterinsurgency, David Kilcullen (2006, 29) writes, is 'a competition with the insurgent for the right and ability to win the hearts, minds and acquiescence of the population'. Winning 'hearts' is explained as 'persuading people their best interests are served by your success' and 'minds' as 'convincing them that you can protect them, and that resisting you is pointless' (Kilcullen 2006, 31). Counterinsurgency is a type of war—'low-intensity' or 'asymmetrical' combat—and a style of warfare that emphasizes intelligence networks, psychological operations (PSYOPS), media manipulation, and finally, security provision and social development that seek to maintain governmental legitimacy (FM3-24 2014). Kristian Williams (2007) makes the distinction between 'hard' (direct) and 'soft' (indirect) practices of counterinsurgency that work in tandem with other larger strategies of population control. 'Hard' techniques—the proverbial 'stick'—include overt political violence by police, military and mercenary forces, while 'soft' techniques—the 'carrot'—are civil-military operations that invest resources and technologies into 'underdeveloped' or 'troubled' areas. 'Soft' interventions are commonly referred to as civilian assistance and community development, including foreign aid provided to collaborating local elites to stabilize and manage areas of interest. Notably, the deployment of these 'hard' and 'soft' techniques often necessitate the establishment of proxy forces or a type of 'decentralized household rule' that work within specific political and cultural contexts (Owens 2015, 8). Within rural Mexico, for instance, this meshes with the (patriarchal) authoritarian corporatism(s) characteristic of many regions (Mackinlay and Otero 2004).

Counterinsurgency takes on a totalizing approach, seeking to monitor and engineer entire populations, and to make them internalize particular values and worldviews. In other words, counterinsurgency is the art of population control, which seeks to colonize and incorporate the people and natural resources into the projects of dominant actors. These are not limited to nation-states and regional governments, but also include transnational resource extraction companies. Indeed, through public or private security forces, counterinsurgency warfare techniques play an integral, if often overlooked component of resource extraction operations of companies (Rosenau et al. 2009; Downey et al. 2010), which sometimes are even calculated into cost-benefit analyses ratios (Caselli and Cunningham 2009). At a

2011 oil conference in Houston Texas, Matt Carmichael, manager of external affairs for Anadarko Petroleum, recommended to public relations experts to '[d]ownload the U.S. Army-slash-Marine Corps Counterinsurgency Manual,' calling opposition to hydraulic fracturing in the US 'an insurgency' (Jarvers 2011). Echoing Carmichael, Matt Pitzarella said: 'We have several former spy ops folks that work for us at Range [Resources] because they're very comfortable in dealing with localized issues and local governments' (Jarvers 2011). The latest counterinsurgency *Field Manual 3–24* (2014, 1–2) explains: 'When a population or groups in a population are willing to fight to change the condition to their favor, using both violent and nonviolent means to affect a change in the prevailing authority, they often initiate an insurgency. An *insurgency* is the organized use of subversion and violence to seize, nullify, or challenge political control of a region'. Counterinsurgency initiatives have led governments as well as resource extraction companies to view civil dissent, non-violent social movements and organized opposition as insurgent or proto-insurgent, where they apply various mixtures of consent and coercion to defuse group formations or their militancy (Dunlap 2014).

COUNTERINSURGENCY IN MEXICO

Mexico had a gruesome history of civilian-targeted low-intensity warfare during the Cold War (Calderón and Cedillo 2012) which re-emerged and reinvigorated visions of rural threat with the 1994 Zapatista rebellion in Chiapas; the 1996 debut of the Popular Revolutionary Army (EPR) in Oaxaca and the subsequent War on Drugs (Norget 2005). The Mexican government responded by restructuring the country around military-centred internal security imperatives of counterinsurgency (Arronte et al. 2000; Stephen 2002). Between the years 1978–1998, no less than 4,172 Mexican military personnel received training overseas, the majority of whom (sixty-one percent) in 1994. Many of those were trained at the US-operated School of the Americas (SOA) and other academies of scientific violence specializing in counterinsurgency, torture and other unsavoury coercion techniques (Arronte et al. 2000). This development coincides with a policy to actively blur the line between the police and military with the creation of the Federal Preventive Police (PFP), in January 1999, who specialize in targeting civil dissent (Arronte et al. 2000). In 1997, Oaxaca State began embracing a comprehensive counterinsurgency plan detailed in a document titled: 'Oaxaca: The Conflict and the Project'. This plan promoted interagency command and control from military, Federal and state police to counter guerrillas, but also to pre-empt the growth of social unrest arising from the 1994 North American Free Trade Agreement (NAFTA) and other government policies (Arronte et al. 2000, 73–80). Summarizing this programme, Arronte et al., (2000, 78) wrote, 'the government's actions in Oaxaca are not geared toward combating poverty, but rather toward counterinsurgency manoeuvres, through the use of publicly funded projects such as the "Microregional Fund", and the media'. Low-intensity warfare operations have now become generalized and reached unprecedented levels under the War on Drugs, which Dawn Paley (2014) argues is actively used to repress civil dissent arising from the unpopular government policies, including efforts to assert control over and further integrate social property into the national economy.

Social property is often perceived by the Mexican government not only as a barrier to economic growth, but also as spaces harbouring illegal activity and potential insurgents (Bryan and Wood 2015). In the *Istmo*, social property—*ejidos* and communal land (*comunales*)—have been a source of conflict for more than a century between and among different tiers of the Mexican government, private companies, land owners and social movements; the most notable being the Coalition of Workers, Peasants, and Students of the Isthmus (COCEI)[6] who eventually came to power in the region in 1981 (see Chapter 1). Originating from Article 27 of the 1917 constitution, *ejidos* provided land for farmers to cultivate, but not to buy and sell—something that began to change after alterations to Article 27 in 1992 (Assies 2008). *Ejidos* are allocated for residential and agricultural use and governed by local assemblies—traditionally constituted by male heads of the household—while 'direct ownership of all natural resources' below *ejido* surface rights were reserved for control by the Mexican Government (MG 2007, 19). Communal land, on the other hand, is related to pre-colonial land claims. Like *ejidos*, it is governed by the community with varying rules and relationships according to regional customs and practices and mostly held in *comunidades agrarias* (Stephen 2002), although internally often treated as private property (Binford 1985). Unlike *ejidos*, communal land has a different legal status that allows a greater degree of autonomy to Indigenous groups over their governing structures and natural resources. Importantly, communal land has served as a barrier towards public and private development projects resulting in a variety of programmes and interventions to better manage and integrate these lands into the political economy of Mexico (Payan and Correa-Cabrera 2014; Correa-Cabrera 2017). In this respect, it is important to recall from Chapter 1 that while 'only about five percent of the land in Mexico was held under communal tenure in 1960," 38% of the Oaxaca land was so administered' (Binford 1985, 180–181).

The trend to officially co-opt and take control of social property began with the 1992 Article 27 revisions that legalized the privatization of social property and introduced the Program for the Certification of Ejido Rights and Titling of House Plots (PROCEDE)[7] that sought to survey, map and register social property (Assies 2008). Lasting until 2005, PROCEDE was met with varying success: one hundred percent coverage in the northern states of Guanajuato and Colima, while predictably less in southern states, with Guerrero certifying sixty-five percent of their social property, Chiapas forty-three percent and Oaxaca thirty-nine percent (Assies 2008). The attraction to PROCEDE was simultaneously intertwined with a 'culture of fear' generated by state agencies to convince individuals that they must have secure titles to guard against inter-communal land grabs (Osborne 2013, 125). Furthermore, it created the prospects to sell, rent, and use the *ejido* as collateral to access farming credit, social welfare and development schemes such as PRONASOL, PROGRESA

(1994–2000) and OPORTUNIDADES (2000–2006) (Assies 2008). However, PROCEDE also sought to collect information on rural communities and create the possibility of privatization (Stephen 2002; Bryan and Wood 2015). This development has intensified with the recent adoption of the Energy and Utility Act (2013) that mandates that social property holders must negotiate and eventually surrender their land to energy companies in regions of development interest (Payan and Correa-Cabrera 2014). While the results of this legislation are only beginning to emerge, analysts see conflict on the horizon (Payan and Correa-Cabrera 2014).

While PROCEDE reached its potential for land titling, classifying and extending government control over natural resources in many areas, the increasing tensions to make social property legible continued with two different, yet mutually reinforcing strategies of what Bryan and Wood (2015, 149) call 'property based approaches to counterinsurgency'. First, the 2006 US military funded *México Indígena,* a 'participatory mapping' project that sought to document native territory and rights to provide land tenure security for residence as well as contribute to the 'open source intelligence' of rural areas around the world (Boyce and Cash 2013). The project was coordinated by Lieutenant Colonel Geoffrey Demarest, led by Peter Herlihy, who managed and supervised a team of geographers from the University of Kansas. This research team set off into the Zapotec *Sierra Juarez* region of Oaxaca to finish what PROCEDE could not, offering mapping jobs, Geographic Information Systems (GIS) training, free computers and updated maps for towns (Bryan and Wood 2015). According to locals, those implementing the project failed to mention that the 'Bowman Expedition' was sponsored by the Fort Leavenworth Foreign Military Studies Office (FMSO), the Mexican Government and the American Geographical Society (AGS) with $2.5 million, with the specific goal 'to gather intelligence on emerging and asymmetric threats to the United States for the purpose of preparing for conflict and maintaining "peace"'[8] (Finn et al. 2014; Boyce and Cash 2013, 245). Through a local intermediary, Gustavo Ramírez, the towns Tiltepec, Yagila and Yagavila began working with the expedition. However, at the start of 2009, once people realized what was being produced and became aware of the military's involvement, individuals and collectives started to denounce the project (Bryan and Wood 2015). Commenting on the 'recent attempts to create world-wide property databases', resident Melquiades Cruz writes,

> this mapping occurs in the midst of the debate over a package of military financing from the United States known as the Mérida Initiative. The control and displacement of Indigenous communities is intended to prevent potential conflicts in 'hot spots', contribute to the military control of the region, and finally free up natural resources for the benefit of the government and its transnational allies (quoted in Bryan and Wood 2015, 144).

In sum, reactions to the participatory mapping project reflect how populations have become acutely aware of the fact that counting, cataloguing and knowledge production are foundational techniques of population and natural resource control.

The second property-based approach to counterinsurgency emerges indirectly with Mexico's national payment for ecosystem services (PES) program. While grassroots movements have adapted and use Mexico's PES programmes to their benefit (McAfee and Shapiro 2010; Shapiro-Garza 2013), arguably PES can serve a secondary purpose of rural pacification. The Mexican National Forestry Commission (CONAFOR)[9] made enrolment in PROCEDE a mandatory prerequisite for participating in PES that creates new real or imagined benefits for farmers (Osborne 2013). However, the related control and financialisation of the natural environment can become easily intertwined with regional counterinsurgency strategies that in Chiapas have worked to blunt the Zapatista movement (Bartra 2007; Osborne 2013). 'In particular state agencies have used conservation and development projects as a kind of counter-insurgency strategy within the Lacandon region', writes Tracy Osborne (2013, 126), making 'participation in forestry programs, such as PES, contingent on the certification of communal lands'. The economic development associated with land certification, expanding roads, plantations and constructing model villages ('Rural Cities') in areas of rural poverty are used to win the 'hearts' and 'minds' of locals while attempting to erode Zapatista territory by confining them within the grid of industrial development and thereby mitigate revolutionary violence (Wilson 2013).

This approach to counterinsurgency, while questionable, is extended with green economy initiatives, not only extending neoliberal logics and markets, but also articulates market-based conservation approaches (PES, REDD+[10]) such as luxury eco-tourism resorts (Agua Azul) and other green schemes. Such 'green' weapons are deployed to advance strategies of land control, population monitoring, industrial development and operationalizing (oblivious/indifferent) tourists as a type of circulating yet continuous settler-colonial force to hold space in contested territory (Bartra 2007; Boas 2011, 47; Osborne 2013; Rocheleau 2015; see also Brock and Dunlap 2018). These textbook counterinsurgency strategies are common throughout Latin America (Bryan and Wood 2015; Grajales 2013; Ybarra 2012; Copeland 2012; Menjívar and Rodríguez 2005; Kohl and Litt 1974) and work in tandem with other such strategies (some of which will be discussed in the next sections examining the development of the Bíi Hioxo wind park).

Green economy initiatives have the potential to not only aid market growth, but also work in accordance with state stabilization strategies that seek to create predictable environments for civil and security sectors to manage rural populations. Even the recent Counterinsurgency Manual (2014, 10–2) openly advocates 'promoting sustainable development', 'education, empowerment and participation' as foundational to mitigating violence and promoting 'long-term regional stability'. Counterinsurgency is designed to be tailored and adapted to local interests, demands and divisions, which are in turn conditioned by larger processes of economic policy, transnational investment, and the federal and state counterinsurgency programmes that shape the practices of locally deployed security services. These larger population management and investment protection strategies trickle down and are adapted into patronage networks, appearing as job opportunities, 'dirty work' or even community development initiatives. Importantly, counterinsurgency aims not only to find social divisions, but also to create and magnify them in order to blur the line between counterinsurgency and inter-communal conflict (Bartra 2007). This is particularly relevant to the existent paramilitarism in Oaxaca and the *Istmo* that is intertwined with the practices and interests of local political elites and industrialists as well as the federal and state security imperatives (Arronte et al. 2000; Stephan 2002; 2013; Norget 2005). Responses to the resistance against the construction of the Bíi Hioxo wind park illustrate these specific dynamics.

FROM DEATH SQUADS TO ARRIVING WIND TURBINES

As the story goes according to people around the Bíi Hioxo wind park, some of the same politicians, industrialist and gunmen[11] involved in putting down the 2006 Oaxaca Insurrection (see Chapter 1) were also instrumental in preparing the way for wind parks in the *Istmo*. These Oaxacan elites recruited people from all over the country for the '*caravanas de la muerte*' (caravans of death), which was a death squad formed to break up and remove the barricades and commune in Oaxaca City (Stephen 2013).[12] It is said that the organizers of this death squad had direct links to the Seventh Sections neighborhood in Juchitán and, after achieving a level of local power, would be assassinated by their handlers, which instigated the intensification of a gang war between two assassin families in the Seventh Section. People interviewed repeatedly claimed that the arrival and security of the wind park was managed by local elites—land owners and politicians. The latest version of the Counterinsurgency Field Manual (FM) 3-24 advises 'to quickly and accurately identify the various community leaders and develop strategies to work with them' (2014, 3– 4). According to interviews in Juchitán, the wind energy companies negotiated with 'political representatives'—formal and informal—paying them anywhere from 17 million to 28 million pesos (if not more) to allow the development of wind energy parks in this region. Exactly how much and where this money was spent remains disputed, unknown and unaccounted for. This, however, hints at the conflict between the civil populations-governments, wind companies-political authorities and between the different political authorities—COCEI, PRD and PRI—collaborating with the wind company in the region.

On 17 June 1964, 68,112 hectares of communal land in Juchitán were officially recognized by presidential decree, which ever since has been a source of struggle and controversy between *comuneros*, land owners, the state and private companies trying to lay claim to the land (Rubin 1997). Currently the issue of communal land outside Juchitán remains unresolved, making the arrival of the Bíi Hioxo project the latest land controversy (CODIGO 2014). The uncertain and actual legal status of communal lands and support of regional elites has led the wind companies to approach *comuneros*—the holders of communal land—to negotiate individual contracts.

The *Bii Hioxo* wind park's approach was to secure land leases through (individual) contract negotiations. According to the CDM documents, before land deals were signed, a consultation process began at the House of Culture in Juchitán in February 2008. People were 'invited to the meeting via loudspeaker in the urban areas where most owners live' with an attendance of approximately '50 owners' (CDM 2012, 43). In September 2009, there was a second meeting, where'[e]ach owner received a personalized invitation at his home', which resulted in approximately 160 owners attending the meeting. Finally, the third meeting took place on 6 December 2011 with personalized invitations for land owners: approximately 155 owners attended (CDM 2012, 44). The language in the CDM document makes clear that the wind company preferred land 'owners', failing to consider the rest of the population in the area or the opposition that had been forming against the wind energy parks. The wind companies and regional political authorities were selective about distributing information about the wind project. Recounting the arrival of the wind park, a seventy-year-old fisherman explains,

> nobody knew about the signing of the contract, they did not ask the community—they did not go house to house like when they are [. . .] campaigning for an election, [when] they go house to house delivering their message. But as for the signing of the [wind park] contract they never got around to asking the people and the people did not give their consent to the construction of that park. Because there was money that the politicians received, but how come upon receiving that money they did not tell the people. So it was built because nobody was asked.

'They arrived here and they said they are going to bring good jobs for the farmers', recounted a farmer who said a wind company engineer tried to pay him 600 pesos for his land. Shocked and insulted by the meagre amount of money offered by the company, he refused—that is 'not enough to feed my family'. Eventually, this engineer offered him 130,000 pesos (approx. USD 6,406) for placing turbines on his land; subsequently, a new engineer was put in place, turning the offer of 130,000 pesos into 3,000 pesos (approx. USD 147). The reason for this change is unknown, but likely a bluff on the part of the wind company. Nevertheless, this particular farmer felt indignation at the thought that the small amount of money would be commensurable with his land. Taking great pride in his work, the farmer saw how the wind turbines would take away his land, pay him next to nothing and ultimately render him dependent on future employment in the city—'They are going to put those things [chains] on our feet again so we work from six [in the morning] to six [in the evening]'. This farmer interpreted this change from his agrarian lifestyle to wind turbine contracts as a step in the direction of slavery to procure his survival. This sentiment was reoccurring as another woman was convinced that the wind company 'has enslaved people through their ambition'. Other people interviewed were infuriated at the prospect of outsiders seizing their lands and disrupting their culture and lifestyle for an insultingly low amount of money.

However, over time people did begin to sell their land. A crucial mechanism that facilitated the acquisition of communal land was the 'public notaries', who worked closely with the wind companies to process land titles. A research participant explained that a notary 'would cost us five to 6,000 pesos per hectare to formalize the paperwork on the lands', but now with the wind company, farmers are filing paper work for ten, twenty, seventy hectares of land for 200 pesos. This was supposedly facilitated by the COCEI, who was acting as an intermediary between the people and state government as the wind energy company was paying the state government to make the land official. The local leftist political party appeared to have had a prior arrangement with the wind companies, who were likely subsidizing notary fees to encourage the titling and privatization of communal land to make it possible to lease the communal land. While some landowners embraced the riches promised from the wind companies, many more remained outraged. This communal land was not only land used for subsistence farming and accessing the sea, but also affected gathering medicinal plants and accessing Zapotec cultural sites.

'The Zapotec religion before the Spaniards arrived; it was a religion that used to worship the sites', explained a social fighter, continuing that 'the sea, the lagoons, swamps, and hills where people would go, for example, to ask for rain. There is the god of rain, the god of corn production; the god of the wind and all of nature to us is something sacred'. When heavy machinery began preparing the way for wind turbines by clearing trees and undergrowth to build new and fortify existing roads, it caused shock and scandal. A woman comments, 'what the business people say is a nasty lie—that it is clean energy, that it is green energy. But how could it be green if it is devastating the trees that provide us with oxygen. It pollutes our ground water. It eliminates alternative natural medicine, which we receive through the animals and plants that are in all the areas where the wind energy companies have invaded'. This conflict came to a head when farmers were restricted from using the communal roads. Taking 'control over public roads, changing the rules of access in the area, and affecting [the] mobility of local people, who find difficulties to reach their agricultural plots' is not only a common tactic employed by mining companies (Aguilar-Støen 2016, 167) but triggered further resentment and direct action against the wind energy company. One research participant explains: 'The farmers would ask the operators of the machinery for permission to go through and they would say, "No. The company has already signed a contract"', which gave construction crews the exclusive right to work there and deny the farmers access to their land. Farmers began organizing, eventually confronting the workers early in the morning. The road workers tried to ignore them and refused to move the road scraper they were operating. Then, according to a protester, 'one of the farmers grabbed his machete [. . .] and cut the [engine] hose in three pieces. And so when the company representative saw [. . .] the attitude of the farmers he sent' for a truck to move the road construction equipment. The farmers waited and were ready to burn down the road scraper until the company brought a truck full of workers to confront them which almost turned into a violent clash.

Eventually, the road scraper was moved, with this particular event serving to raise awareness about the numerous and myriad changes created by wind park development.

The takeover of communal land had serious cultural implications. Not only is the communal land the lifeblood of farmers, but it is also home to seven sacred sites where people embark on a pilgrimages to pray, to keep in contact with nature, to relax, to socialize and to share the harvest of the year with each other. Among these are important historical sites where the Zapotecs resisted the Spanish: the tombs in Guze Venda, Guela Venge, Chigueze and the 12th of May in La Chaxada, in Paso Cruz.[13] A women explains,

> When you enter Guela Venge, now the main entrance to that chapel is blocked by the wind energy project. They no longer let people to freely enter there. There are people that go there to pray, there are spiritual sisters that go to do their rights there and they are no longer allowed to enter in that area of Guela Venge. In Guce Venda, because of the resistance of the APPJ [resistance organization], people can still enter there, but the company had intended to close that off. Why? Because they started to block near Chigueze, they did not want farmers who had not signed on with the company to go through there. That is where the APPJ started. Yes, because they were farmers and fishermen who wanted to go through there and they were told that they could go through, but only with a badge on.

Then, armed security guards patrolling and preforming identity checks on the communal roads were transforming the land relations in the area and generating resentment. Residents were effectively treated as prisoners on their own lands. Complaints surfaced about how the pilgrimage to the Holy Cross of Chigueze[14] was not what it once was. During the pilgrimage 'a prayer is done at each farm and people share their produce with their fellow countryman [where they] would go from farm to farm until they arrived at the shore of the lagoon'. This year, however, people did not share on surrounding farms. Whether this is the result of fences, security guards, people not producing as much food, or the wind company discouraging their signatories from supporting the pilgrimage is unclear, but the cultural change was taking place. The Chigueze chapel sits about 400 meters from a wind turbine and now the communal road is lined with security cameras (see figure 3.1); meanwhile, the road is patrolled by heavily armed private security guards (PABIC). Wind companies are employing the same tactics as other extraction industries in Central America by seeking to 'prohibit or impede community member's access to their farmlands using armed guards, fences or locked gates' that serve to provoke systematic discomfort and 'to force peasants to sell their lands' (Aguilar-Støen 2016, 165). Wind park development is slowly aiding a wider tread of cultural erosion and further industrial integration, which is triggered by forced (as well as voluntary) communal lands changes that have been deeply tied to Zapotec cultural life for centuries in the *Istmo*. For the people who valued the land, sea and their culture, this situation has left them little choice but to defend their land and sea at all costs.

Figure 3.1. Security camera along the communal road in the Bíi Hioxo wind park.

Before the formation of the Popular Assembly of the Juchiteco People (APPJ) to halt the arrival of wind turbine equipment with the barricades described above, people tried to stop the construction of the park in other ways. Confirming the point made by a research participant, I asked the question: 'So essentially, what you are telling me is that this wind park was built here with people guarding it with assault rifles while it was being built?' The person replied:

Yes! For example over there [pointing], we came and chased them, but because they have weapons, they started shooting at us—*'Bow! Bow!—Bow! Bow! Bow!'*—and one of the bullets hit really close to me and we started running, but they took a kid, and a lady that came with us and all the women that were strong started yelling at them: *'Release him! We did not come to fight with weapons, we just came to try and stop these wind turbines from being here.'* And one of them [gunmen] took his pistol and started shooting—*'Pow! Pow! Pow!'*—but he didn't hit anything and he ran out of bullets. So the other guy handed him a knife and started saying, *'Kill him! Kill him!'* [holding the kid], but he didn't want to and the other one kept yelling, *'Kill him! Kill him now! Kill him!'* So the kid was really scared, but he was really lucky that the women took him back, they took him from those guys' hands, but he ended up with his t-shirt ripped.

AD: *How did the women grab him from the gunmen?*
Pulling him. One of the women started yelling, *'Kill me and you will realize that we do not have weapons, but you are not going to take this kid anywhere to kill him.'* So we stayed away [from the gunmen] and nobody was allowed to pass [to the communal land]. We made a barricade, and they paid the gunmen to burn it.

The loss of the communal roads produced the barricade, and the ensuing battle (described in the introduction) subsequently required state police to escort wind turbine construction equipment onto communal land. The limitations of resistance were felt when the construction workers of the wind company began applauding the barricades as they resulted in pay raises for them, without work. However, the legal battle and resistance in Juchitán, as well as San Dionisio del Mar, Álvaro Obregón among others, was able to summon a Free, Prior and Informed Consent (FPIC) inquiry that will be analyzed in Chapter 5. Resistance in general created space to question these projects making implementation more difficult, meanwhile creating the possibility for negotiating better benefit-sharing for individuals and towns. That said, resistance generated a violent response from Gas Natural Fenosa, Oaxaca state and local governments that engaged in a concerted campaign to undermine the resistance against the construction of the Bíi Hioxo wind park. Revisiting counterinsurgency techniques, the following section will examine some of the tactics and strategies used to facilitate the construction of the Bíi Hioxo wind park on communal land.

DIVIDE AND CONQUER: COUNTERINSURGENCY FOR WIND ENERGY

The strength and militancy of the anti-wind energy social movement in Juchitán and other parts of the *Istmo* no doubt constitutes a threat to existing and future wind energy development. This section reviews the 'hard' and 'soft' counterinsurgency techniques applied to overcome this resistance by Gas Natural Fenosa (GNF) and their collaborators within the Mexican government, regional elites (*caciques*) and their networks.

Hard Techniques

The security forces operating in Juchitán are diverse. There is Military, Naval and Federal police presence combined with state, municipal and the Auxiliary Bank, Industrial Presence and Commercial Police (PABIC). At the same time, there are roughly three types of mercenaries: (1) local gunmen, who know or in some cases are related to the people struggling against wind turbines; (2) Mexican gunmen from other parts of the country; and (3) gunmen that look like foreigners—'whites'—that are rumored mercenaries hired by the wind companies. In the struggle against the Bíi Hioxo wind park a variety of repressive techniques have been documented by the *Comité de Defensa Integral de Derechos Humanos Gobixha* (CODIGODH 2014, 26–8):

- Police harassment and brutality
- Death threats (in person and by phone)
- Home Surveillance
- Firing guns in front of homes
- Breaking into and vandalizing homes and community radio stations
- Creating informant networks
- Attempted kidnappings
- Illegal detainment (held hostage) by police
- People being followed around town and after public events
- Burning individual ranch lands
- Burning the road blockade camp (three times)
- Assassination

On 19 May 2013 a farmer was shot working his field, while on 21 July 2013, after he left a party, APPJ member Hector Regalado Jiménez was shot six times. Badly wounded, he was rushed to a hospital where he identified two men and a member of the PABIC as the perpetrators, thereafter passing away on 1 August (CODIGODH, 2014). Many of these attacks were documented by the *Comité de Defensa Integral de Derechos Humanos Gobixha* (CODIGODH 2014, 26–28), but their list is only a partial chronology judging by the recorded interviews. Many attackers are well known 'because sometimes the people who threaten us are our families or cousins at the service of *Pulpo*'. Questioning the appearance of white mercenaries, a man living inside the Bíi Hioxo wind park replied:

> Yes, there are white people. I have seen one that always goes at the head of the caravan of vehicles. He is white, tall, light colored eyes and a straight nose—kind of like you. He is always at the front of the caravan and everybody else is behind him. So since I am involved in the opposition group and I am visible when our group gathers, when I would go out to lead my cattle to graze and give them water people would shoot at me. It is a threat—they were trying to intimidate me.

AD: *Did they shoot in the air or at you?*
At me, I had to run. I know all of the land around here.

AD: *How many times has that happened?*
[. . . .]
Four times.

Reports of shootings are frequent and the details of those managing these types of coercive actions are common knowledge. The inaction on part of the authorities, one could only guess, is likely because of the support the companies receive on a federal, state and local level, with political elites, politicians and their supporters collaborating with the efforts of the wind companies on various levels. Despite all efforts to the contrary, usually packaged in drug war legislation and counterinsurgency aid (Paley 2014), systemic coercion is normalized in Mexico and constitutes its own realpolitik intimately tied to economic interests in the region (Correa-Cabrera 2012).

The wind company also recruited gunmen from the barricade. 'The wind energy companies sucked them in with money and they armed them and turned them against us', explains another fisherman who said they harass him when he is using the communal roads. 'They use killers from this neighborhood and they also use them to tear apart our assembly—they use our very own *compañerxs* that used to be part of the assembly [APPJ] and then bribed them with money', explained another man who continued, 'among them is one of my uncles—a brother of my father—who has also received money from the [wind] companies and with that took some *compañarxs* with him'. Summarizing, a woman explains: 'they are given money, they are given weapons, and they do not have to work hard, so they join the company'. Buying and flipping the loyalties of people became a problem for people in resistance.

Similarly, as should be expected, police functionaries and informants were used. They mined information, stole community radio equipment, spread lies and exaggerations about people in the APPJ. Notably, about two months into my research, someone began spreading rumours that my translator and I were police informants. According to the rumour, which referred to our descriptions rather than our names, information from Chiapas said that we were infiltrators. In this area, and others experiencing social conflict, social scientists, journalists and NGO personnel remain questionable figures as they have come to serve a vital role for militaries, police and resource extraction companies by providing direct information or indirect 'open source intelligence' (as discussed above). Collecting information on the people, their opinions and dispositions serves to help governing authorities or extraction companies create strategies to undermine opposition to their policies and projects. The lies spread against me and my friend served a different function which sought to undermine my relationships with the people fighting the wind energy parks. Later, I was informed that it was a 'leak' from inside the Salina Cruz courthouse that informed people in the resistance that I was a police informant. Meanwhile, I was being harassed, and six people from a state-sanctioned mercenary group in Álvaro Obregón calling themselves Constitutionalists tried to grab me.[15] There were also accounts of two female social scientists working with Mareña Renovables[16] in San Dionisio del Mar, which the community, according to the research participant, eventually pushed out of town. The women responded by exclaiming that the company would return the following year or year after. Information gathering was recorded taking on a more clandestine approach. A land defender in Juchitán explained how human rights organizations were used to gather information 'to look who the leaders are', proclaiming themselves 'as human rights defenders or members of collectives or from some university'. Information gathering using the internet, NGOs and community development is textbook counterinsurgency operations in both rural and urban areas (FM 3-24 2014; Williams et al. 2013; Kilcullen 2006), which seeks not only to collect information on opposition groups, but also to divide people through distrust, paranoia and create an overall climate of insecurity (Williams et al. 2013). This creates stress and dissolves friendships, families and support networks, creating an extended cost to opposing the government and their business affiliates—in this case wind companies. While these repressive tactics were being executed, the Bíi Hioxo wind park was simultaneously articulating 'soft' counterinsurgency techniques to fragment and break opposition to the project.

'Soft' Techniques

The campaign waged to integrate the Bíi Hioxo wind park into Zapotec society is best described by the section titled 'Integrated Monetary Shaping Operations' (IMSO) in the chapter 'Indirect Methods for Countering Insurgencies' in FM 3-24 (2014). IMSO is the use of money to socially engineer an environment to accomplish the goals of state security forces or private companies. After the battle of the barricade, a study was conducted titled *Perceptions of the Wind Energy Park in Juchitán, Oaxaca*[17] where, according to the study: 'Gas Natural Fenosa needs to implement marketing strategies to understand the Juchitán, Oaxaca population's perception of the Parque Eólico Fuerza y Energía Bíi Hioxo *with the goal of introducing possible solutions to deactivate the social movements that have arisen around this project*' (emphasis added). Presented in June 2013, this study was based on 225 phone interviews and sought to find ways to make the installation of the Bíi Hioxo wind park politically feasible. The following is a translation from the 'Conclusions and Recommendations' section:

- There is an observably high level of knowledge about the wind energy park, but there is not a perceptible homogeneous opinion about the social and economic impact in the area of influence.
- Although there are clearly identified groups opposed to the project, no clear rejection to the project was found, but neither was a decisive support of the creation of the wind energy park; which represents an area of opportunity for changing its position in the public opinion.
- The main benefit of wind energy parks that the Juchitán population perceives is the creation of employment.
- This expectation is based more on the hope/desire than the certainty that the creation of employment will happen.
- Collective welfare is a desire-attitude that determines the interaction of Juchitecos [with the project]; a project should be discursively addressed in plural form to promote collective appropriation.

Hot Spots

- Within the current discourse the position that is managed by the opposition is stronger because it centres on the personal safety and the economic, social and natural heritage of Juchitecos—it is worth emphasizing this town's almost religious connection with nature and its environment.

- The youth are observably the group least informed about the project, but with greater credibility in social networks, these media are an important avenue of approach to this segment of the population.
- There is fertile terrain in which the opposition can take advantage, due to the lack of strength in the institutional discourse that supports the project. In this study one perceives that it is possible to more solidly communicate the benefits of the project given that nine out of ten interviewees show a readiness to support it.
- There is a notable climate of speculation that should not be overlooked, which is the anxiety awakened by this kind of project that can, given the timely and clear dissemination of information, turn into a time bomb—which is to say it is a potentially dysfunctional event which must therefore be strategically deactivated.

While the study was not as comprehensive as *México Indígena*, these two Power Point presentation slides demonstrated how even rudimentary phone studies and 'opinions' could provide enough information on the human terrain in an area to help identify the next actions that should be taken, or be avoided, to complete the Bíi Hioxo wind park. This study recognized the strength of the resistance as well: '69% of those who know the group [APPJ] are against the project, while 34% give it the benefit of the doubt' (p. 27). It established the importance of targeting youth groups, identified a 'desire-attitude' within the population to manipulate, and found the people would still listen, which emphasized the importance of forcefully and strategically disseminating information supporting the wind park—otherwise it could be 'a time bomb'. This is reminiscent of Kilcullen's (2006, 33) conflict management advice (i.e., to administer 'a simple, unifying, easily-expressed story or explanation that organizes people's experience and provides a framework for understanding events'). Gas Natural Fenosa operationalized this information by further developing supportive fishermen groups, social development projects and public relations campaigns that cumulatively instigated social divisions, wind park acceptance and cultural erosion.

The way people were bought off on the barricade was also replicated with fishermen. The wind companies organized a cadre of fishermen to attend public events and speak in support of the Bíi Hioxo wind park. There are two principal groups of cadre fishermen,[18] with around 250 members supporting wind parks in the area. The Secretary of Indigenous Affairs provides them with work through a 'Temporary Work' program as well as fishing equipment: boats, motors, nets, weights and fishing line. However, many of these people are not fishermen or do not fish anymore. According to people in interviews, these fishermen go to towns on the Laguna Superior and sell their equipment worth around 800 pesos new for 200 pesos.[19] People resisting wind projects feel these fishermen groups, who are openly supporting the wind companies for money, were created specifically to counter their opposition to the wind park—counter-fishing groups were created, as opposed to counter-terror groups advocated by counterinsurgency theory (Kitson 2010/1971).

Social development projects were also used to discredit the ideology of 'the opposition'. While some people have always been negotiating with the wind company for social development, these projects began manifesting themselves in 2013 and 2014. Interestingly, the Bíi Hioxo wind park used what had traditionally been activist counter-information tactics to defuse oppositions groups (Juris et al. 2008). This came in the form of the WordPress website 'Parque Eólico Bíi Hioxo' that was dedicated to promoting and countering information against the wind park. The website section titled: 'Myths and Realities' asserts that the wind park 'is environmentally friendly;' 'The lands where wind farms are installed are still usable' for 'livestock, agriculture, fishing, etc.;' 'It is not Loud;' 'It is safe;' and finally, '[I]nvolves Economic benefits for Communities' (GNF 2013b). As we will see with local accounts in the next section, these claims are questionable.

This blog also documented and reposted all of the positive press and contributions made by the wind company to the surrounding communities (GNF 2013e). Points six and eight of IMSO that stresses the importance of '[p]urchasing education supplies', 'providing education to the local population' and '[r]eparing civic and cultural sites' as 'cultural heritage is a sensitive issue' (FM 2-24 2014, 10–12). Here you can see an adherence of the Bíi Hioxo wind park to these points by providing computers for schools, medicines, household appliances, food supplies, repairing cultural sites, conducting a reforestation campaign, fixing an irrigation canal, and providing electrical engineering classes at the Technical Institute of the Istmo. Additionally, the wind company tried to help with church restoration and get involved in local customs, while systematically denying access to cultural sites. Overall, despite the attempts to curry favour, these monetary shaping operations still generated outrage among residents. This manifested when teachers brought their students to participate in the Bíi Hioxo sponsored reforestation program, which was protested and refused by some students. One person explained that the boy told his teacher that the wind companies are 'fooling the people and stealing their land', while another girl when they were trying to force her to 'plant a tree and take a photo of her, said, "No! They are fooling the people and the kids this way"'. The kids continued to explain that the photos were being used to show that the wind company 'is helping people, but it really is not true'. While some kids protested, others complied rather than challenge the authority of their teachers who were working for recipient schools of wind company social development aid.

These social development programs were intertwined with other public relations campaigns. FM 3-24 (2014, 3–4) states: 'Stories, sayings and even poetry can reveal cultural narratives, the shared explanation of why the world is a certain way. Frequently, advertising appeals to people by using these narratives, as do effective information operations'. Notable among these is the *Winds of Change* photo exhibition by 'local artist' Jacciel Morales (GNF 2013c). This was a photo exhibition that tried to blend Zapotec cultural symbols with wind turbines, attempting to send the message that the two could co-exist (see figure 3.2). The pictures show cows under turbines; a woman in traditional dress playing with her kid with wind turbines in the background; as well as a farmer reflecting into the sunset with wind turbines. Again, some kids knew better.

> They set up a photo exhibition in the house of culture. In the schools that have received computers and paints to paint the school they told the teachers that they need the kids to come to the exhibition. So they dragged several kids to the exhibition. Among those kids was the daughter of Juanito and this girl told the teacher that they did not want to go to the exhibition because they are fooling the kids. So the teacher said: 'If you do not go, you are going to lose points for your grades.' And the girl said: 'I do not care I am going to tell my mom.' So Juanito's wife, Maria went and told the [other] parents that the kids were going to be taken out of school without the parents' permission to go to the exhibition, with the intention of fooling the kids to convince them that the wind turbines do not pollute and that people can live with them. So by the time the girl had told her mom, two buses have already left the school paid by the [wind] company and they took the kids without authorization. So from this incident a group of women was formed, among them Juanito's wife, and they went to the House of Culture and they yelled at the principal: 'Hey you are fooling the kids! And that is in violation of children's rights! And you are also going over the parent's authority.' So they yelled and the kids heard and the principle called the state police.

Another person commenting on this attempt to fuse Zapotec culture with wind turbines:

> they started to get involved with people who worked in the cultural area—writers—they published books written by some poets from this region and the books would have the logo of the company on them and then they could say that they were supporting our culture, our art, our way of being and things like that—a way of fooling people. By publishing a book they could say that they were involved in the community. A series of things—lies—but they were just crumbs in order to take control of our territory.

Consistent with the IMSO and culture section in FM 3-24 (2014), these public relations campaigns also included raffling off household appliances, buying trophies for soccer tournaments, distributing Bíi Hioxo windbreakers to kids in the winter, and supporting a breast cancer fundraiser. Nevertheless, these impositions went deeper than public relations and began sowing divisions within Zapotec religious groups.

While Zapotec appropriation of western religions under colonialism is well-known and religious divisions are common (Tutino 1993), the wind company began donating money and resources to churches to intensify divisions within religious congregations. These internal divisions were compounded by media lies about the people in resistance, a popular one being that the APPJ is 'receiving money from the [wind] companies'. Trying to get involved in local cultural events, the wind company also began sponsoring counter-*Velas*. *Velas* are large, often extravagant day—or even week-long cycles of festivals that take place in and around Juchitán. They are often

Figure 3.2. Photo from the Winds of Change Exhibition

centred on processions or a mass with food sharing, ritualized dancing and drinking (Rubin 1997). While the wind company had already blocked communal roads, damaged and altered cultural sites, they have 'created an imitation of the three crosses and looked for a new Majordomo who was financed by the wind energy companies to make their own Vela', explained a Zapotec priest who continued:

> We made the Vela on 30 April and on the same date they made their own Vela in the northern part of the city. We looked for a music group for a moderate price and they [the wind company] looked for expensive music groups and they brought two of the best groups from Oaxaca City, so they could draw more attention to them and attract more people. They gave away free beer, food, snacks and everything. And during the pilgrimage we brought the cross with us while we walked towards the chapel, but instead of doing the same, they took the cross to the northern part of the city to the house of the Mayordoma.

Fuerza y Energia Bii Hioxo's 'firm commitment to respect for human rights and specifically the traditional ways of life' created new religious divisions in the area, causing discord, discontent and social fragmentation. By giving money to people in traditional Zapotec-Catholic religious groups, they found ways to create divisions and widen their support base.

Hence, corporate social responsibility and their claim to respect 'traditional ways of life' and 'human rights' could simultaneously work to further the strategies of the integrated monetary shaping operation (IMSO). Social development initiatives can serve to isolate those resisting the wind energy project, systematically refusing money and negotiating the coercive tactics deployed by the wind company and their local collaborators. On a personal level, these techniques have resulted in sickness, poverty, loss of friends, fragmentation of families, and paranoia among other discomforts. 'El

Jefe' explains how, because he took a public stance against the wind turbines, people at his work see him 'as a weirdo and they do not want to be around me anymore because they say, "Nah, the law is following him."' This extends to people turning down rides, colleagues joking: 'when are they [the police] going to arrive and arrest you?' and the overall construction of an environment designed to imbue apathy and submission to 'business as usual'. This strategy tends to create isolation, feelings of rejection and combined with coercive techniques, to stress that can lead to health problems. Explaining the importance of his wife, kids and friends to undermine these isolating tactics, El Jefe gives an example of how this plays out in daily life:

> [A]t the end of the school day we had to split up, my kids and my wife have to go one way and me another, while also being very careful if cars or somebody come really close and I told them: "If that happens, run and ask for help." This is why this is creating for us some kinds of sicknesses with my family, my brothers and my sisters.

'The Wild Tiger' elaborates on the impact of this repression: 'Rights such as the freedom of movement, freedom to socialize and freedom to enjoy public space—you lose that. You sacrifice that in order to defend the land and territory'. This includes not attending the Velas. Every facet of cultural life: relationships with the land, the sea and their cultural practices—the very foundations of indigenous life in the *Istmo* are undermined by these subtle, yet divisive practices deployed in a context of material poverty that seek to buy the hearts and minds of people. The politicians and *caciques* funded by the wind company would intervene not only using a mercenary apparatus, but also promising riches and distributing money, finding willing participants to support and by extension divide—strategically—every facet of coastal Istmeños life to build the first wind park on the Laguna Superior.

BÍI HIOXO REALIZED

Facilitated by money, mercenaries and police that applied textbook counterinsurgency techniques, the Bíi Hioxo wind park became operational in October 2014. While resistance was fierce and continues today, the trauma of both 'hard' and 'soft' psychological operations, violent attacks, murder, and clashes with state police eventually enabled the wind park construction. Not long after the wind turbines were built, reports of negative environmental impacts became frequent—a critical topic in the FPIC consultation discussed in Chapter 5.

During research, fishermen repeatedly told me how wind turbine tower lights and vibrations severely impact fish, pushing fish populations further into the Laguna Superior and therefore preventing the common practice of fishing on foot in the Laguna's shallow waters. Similarly, there are reports of large quantities of fish dying from wind turbine test-installation on the Barra de Santa Teresa near Álvaro Obregón (Chapter 4). These developments triggered social conflict between villages. A number of the fishermen who signed on with the Bíi Hioxo wind park in Playa

Vicente and Juchitán started travelling to fish in neighbouring villages who were struggling against the invasion of wind turbines—'[T]hey work to destroy the sea, but they still take from it'. This creates a division *between* towns, which are already divided internally over the issue of wind energy development. A member of a local human rights group described this as an 'inter-ethnic conflict because of the wind energy projects'.

Road and wind turbine construction necessitates the transformation of land, the clearing of trees and brambles that results in habitat loss for birds, iguanas, armadillos, rabbits and cougars, among others. This not only results in animal deaths over a long period of time, but also the loss of biodiversity and plant species important to local medicine.

> They are harming our oxygen, our lagoon—the lagoons give us shrimp, crab, fish—and the community lives from this, especially in the Seventh Section [. . . .] Like[wise], there are over two hundred types of medicinal plants and each medicinal plant has its area, its natural habitat. So, when the wind energy projects invade, they kill this area, they kill part of the herbs. How are they going to transplant those herbs to another place that is not their natural habitat?

Hence the logic inherent in 'offsetting' (Sullivan 2009; 2010; 2017), which neglects the sensitive, context-specific needs of local species for development projects that transport native plants to new locations. These concerns over animal habitat, fishing, hunting and livelihood are reiterated frequently and are the driving force behind the APPJ and others who are resisting the Bíi Hioxo wind park and other wind turbine projects in the *Istmo*.

Clearing habitat for roads is compounded by wind turbine foundations. Wind turbine foundations have intense impacts on the hydrology of ecosystems, triggering both extreme drying and flooding as the concrete prevents drainage into the soil and the Laguna Superior (see Tabassum-Abbasi et al. 2014). The communal land is the wettest area in the *Istmo*, with fresh water 1–2 metres below ground, and it is transformed into concrete for wind turbine foundations. Protected by heavily armed mercenaries—'they were working day and night', explained a farmer who continues by describing the foundation construction process:

> 'They brought a lot of machinery, those for digging, they made a ravine and a *square that was 20 X 20 metres and it was 12–15 metres deep*. So for example where that one is standing [there . . . .] *After they brought some fluids and they poured them into the water and I do not know what happened, but after that the water stopped*. They were working really fast; they brought over fifty dump trucks full with stones so they can make roads. That is how they made the highway so fast. They were putting down stones and then another machine that has a blade was levelling and another one that was rolling to compact the stones. They worked really fast.

The wind turbines have turned the aquifers into concrete, and these impacts are compounded (literally) by road construction. '[S]ixty centimetres all around my field' said the same farmer, explaining that when it rains, his land is turned into 'a pool'. Before the wind park, water used to drain 'into the ocean', but now it collects on his land and turns it into a pool. This farmer is resisting the placement of wind turbines on his land, even while he is now surrounded by wind turbines as close to thirty metres to his rancho. This, again, is a common tactic of extraction companies, where 'the fields of people who refuse to sell their land become engulfed within large areas owned by the companies' (Aguilar-Støen 2016, 165). Likewise, according to this farmer among others in the region, the roads, foundations and wind turbines cause extreme drying and flooding of crops. During the rainy season his watermelons float in over a foot of water and rot. The excess water is combined with oil leaking into the ground. Oil coming off the turbine drips into or seeps into the water wells. This same farmer explains: 'The wind turbines spit oil on some of the land; and then the machinery comes, and they dig and lift the earth [to] make a pit; and all of the land they moved, they throw it into a trench, so it doesn't affect the land anymore'.

People spoke frequently about the noise, and how it made them feel sick, which was also a problem in La Ventosa (see Chapter 2). Another farmer said he was going to move out of his house because of the noise hurting his eardrums, giving him vertigo and chills, and making his brain feel like it 'is moving and contracting, and that makes me nervous'. Wind turbines 'provoke headaches and dizziness, and this is an illness either of the nervous system or hypertension', a local herbalist tells me, continuing that 'before you went to a ranch . . . you hear the birds sing, the growling of the wind, all of this relaxes you, but now it is not that way. You go there to the rancho and you start hearing a bothersome *buzzzzzzz*—how can you relax?' The land relationship is completely altered, from animal noises to electrical power lines and grinding wind turbine gears. For people living with and connected to the land, the impact of industrial wind turbines is significant and by no means 'green' or 'clean' energy and could only be conceived in these terms by comparing wind energy to the annihilating processes of coal, oil and nuclear energy production.

CONCLUSION

The impact of industrial wind turbines has been overwhelming: creating social divisions, violent conflict and altering the natural environment to a point of destruction. Despite widespread popular protest, the outcome of the Bíi Hioxo wind park was facilitated by counterinsurgency warfare techniques deployed with implicit and explicit support of public and private institutions to articulate a campaign of pacification to install the wind park on communal land. This application of coercive force displays the forceful and disingenuous nature of wind energy development, the green economy and climate change mitigation practices in Mexico that are entangled with the growth imperatives of states and their national and transnational business affiliates.

It is true that wind turbines do not generate 'radioactive waste' as the CDM document mentioned above states; that is, unless it is a direct-drive wind turbine that can generate radioactive waste when mining rare earth minerals for their permanent magnets (Dunlap 2018). The Bíi Hioxo wind park uses geared or doubly fed induction generator (DFIG) and does not generate radioactive waste (Gamesa 2018), wind turbines do, however, require 'extraction, drilling, [and] transport of fuel' so oil can lubricate its propellers, metal towers can be constructed and rare earth minerals can be processed for their magnets (Hoenderdall et al. 2013; Guezuraga et al. 2012). The true cost of wind energy tends to be concealed in other energy intensive fossil fuel based infrastructures, which is compounded by the location of the Bíi Hioxo Wind Park in an area rich in fresh water resources, biological diversity and animal populations directly tied to the subsistence activities of Zapotec and Ikoots populations. The spread of wind energy serves to expand industrial degradation, first in the construction of wind turbines themselves and secondly by energizing the image and operations of natural gas companies, superstore chains and industrial-scale construction companies that are expanding their business operations using wind energy resources. This means the location, placement, quantity of wind turbines, and, most importantly, how the energy is used will determine the impact and ecological sustainability of wind parks. Is wind energy being used to cut back consumption and transition away from degrading extractive industries and consumption patterns, or is it instead being used to expand, entrench and intensify patterns of capitalist consumption and infrastructural expansion? This question remains important when assessing the development of 'green', 'clean', or 'sustainable' development projects.

The environmental impact of wind energy is intimately intertwined with land politics, economic growth, state and extra-judicial violence to implement, manage and secure project investments. Consequently, the Bíi Hioxo wind park serves to renew the image of Mexico as a leader in environmental policies, develop a wind energy market and acquire Indigenous land in the name of climate change mitigation strategies. The green economy thus serves to advance resource control and environmental degradation, supporting Dunlap and Fairhead's (2014, 954) assertion that the green economy is 'a continuation of war by ecological crisis'. This war is not lost on those in resistance against the Bíi Hioxo wind park. 'The first nations have contributed to the protection of Mother Earth, but now for those who have money we are an obstacle,' said a Zapotec land defender who continues by saying that businessmen are

> taking advantage of Mother Earth's illness, which is global warming, so they can grab all the natural resources from the first nation people from this land. They are grabbing our water, they are grabbing our wind, they are grabbing our lands, they are grabbing our forests, and they use the protection of natural resources as an excuse. [. . .] [T]he responsibility of those with money—the corporations that are polluting the land, air, and water. Then they use all their actions that caused Mother Earth to get sick as an excuse to make us pay for their damages.

In the context of climate change mitigation deliberations, for people struggling against wind energy projects the responsibility lies with the corporations and the industrial system of environmental devastation—in one word: capitalism. For those in resistance, 'the wind energy project for us is a tool of ethnic cleansing that does not build anything useful for me'. The words of Owens (2015, 249) are telling in this situation: 'modernisation theory provided the distinctly liberal and progressive rationale for crimes against humanity', which we can see have now been morally armoured with 'green' and sustainability discourse. The anthropology and sociology of genocide (Short 2010; 2016; Levene and Conversi 2014; Moses 2010/2008; Hinton 2002) have already shown the vulnerable nature of Indigenous cultures surviving colonialism and facing what amounts to 'a slow industrial genocide' in the way development projects are poisoning and degrading the land base of Indigenous populations (Huseman and Short 2012).

In the *Istmo*, not only are wind parks imposed by counterinsurgency techniques—but wind turbines themselves destroy animal habitat; poison the land with chemicals, concrete, and oil (see Chapter 2); kill, and displace aquatic life into the Laguna; and fill the air with a constant 'Buzzzzzzz' from electrical currents twenty-four hours a day. Wind turbines in the *Istmo* are a direct attack against the land base of Zapotec, Ikoots and other people living from the land, which can be viewed as an indirect starvation strategy deployed in the name of neoliberal environmentalism, in order to deliberately remove the means of survival from those in areas of key investment. Starvation strategies are a classic counterinsurgency technique, and are frequently deployed against nomads and Indigenous populations across the world and over the centuries (Levene 2008; Boot 2013). The US military is well known for targeting the crops, sheep, fish and, most famously, the near eradication of all buffalo herds to undermine Native North American resistance to the expanding US frontier (Isenberg 2000; Churchill 2001; Boot 2013), which gave way to reservation resettlement and later assimilation campaigns to integrate Indigenous populations into labour markets, export-oriented economies and the political culture of the United States (Moses 2010; Wolfe 2011).

In this sense, wind turbines represent a modernized or 'green' starvation tactic that destroys subsistence activities, land relations, and, by extension, Zapotec and Ikoot culture, which indirectly displaces these groups into greater dependence on the national and global economy, contributing to the creation of a reserve army of migrant labour. Arguably, relationships with the land are being substituted and further entrenched by economic processes that will further degrade it. Megaprojects—wind or otherwise—that degrade the land and poorly compensate entire populations, cause green grabbing to flirt with or become 'green extermination' or assimilation depending on its intensity. This is done implicitly by the imperatives of states and their economies, whose intention is to legislate and enforce, often drawing on counterinsurgency techniques, accumulation strategies that further integrate people into a political economy bent on ecological destruction. This raises the question of 'cultural genocide' that will be revisited and discussed at length in Chapter 6. Before this, however, the book will turn to the wind energy conflict in Álvaro Obregón, which is followed by Chapter 5 documenting and analysing the first Free, Prior and Informed Consent (FPIC) consultation in Juchitán.

NOTES

1. Asamblea Popular del Pueblo Juchiteco.
2. A scarf that is part of traditional dress in the *Istmo*.
3. A cloth pouch.
4. See promotional to view uses of wind energy: https://www.youtube.com/watch?v=3M1hs5r6ws.
5. This is a voluntary initiative based on 'CEO commitments to implement universal sustainability principles and to take steps to support UN goals' (UNGC).
6. *Coalición Obrera, Campesina, Estudiantil del Istmo*.
7. *Programa de Certificación de Derechos Ejidales y Titulación de Solares*.
8. See Dunlap (2014) for a brief etymology of 'peace'.
9. Comisión Nacional Forestal.
10. **R**educe **e**missions from **d**eforestation and forest **d**egradation, and foster conservation, sustainable management of forests, and enhancement of forest carbon stocks.
11. This is well-known in the *Istmo*; names will not be used here and come from an undisclosed source.
12. The people's work also included organizing Movimiento de Unificación de las luchas Triquis (MULT) paramilitary groups in the Sierra Sur and Mixteca, which intensified divisions and killing among Triqui communities. See Stephen 2000; Norget 2005; Gibler 2009 for more on this struggle.
13. The spelling of Zapotec words tends to vary and some of these spellings were not confirmed.
14. Brief etymology: 'It is a very old Zapotec word. At this point it cannot be defined. There are some people who say that it is a thorn, it is derived from gichi and cueze is some of the properties of the thorn. There is another term, Chi, which means they and 'Ic gue che' to us means priest. So there is no definition because it is an ancient word, it is more than a thousand years old—there is no proper definition. In the same way the definition of Igú, many say that is a sweet potato, but I am telling you the elements of the ancient era were considered sacred'.
15. Commonly known as *Los Contras*.
16. Now Eólica del Sur.
17. Percepción de Parquet Eólico Juchitán, Oaxaca.
18. Supposedly led by Mariano Asuncion Castillo and Roberto Castillo.
19. 'This guy came and sold me a package of weights and fishing mesh and I know he is not a fisherman because he drives a Moto taxi', recounts a friend who continues by saying 'for example, if a kilo of weights is worth 200 pesos, they are selling them for 50 or 40 peso a kilo'.

4

Insurrection for Land, Sea and Generational Integrity in Álvaro Obregón

> Totopo is the Resistance
>
> —Mr. X

Invited into a backyard shaded by Tamarind trees, the human rights defenders pull up chairs. We sit down, and one defender picks up a stick and begins drawing a map of the coastal *Istmo* in the loose dirt on the ground. Telling the history of the wind energy development in the region, he explains that 'there is a 1,000 year old cultural identity and it has a lot to do with the sea, with the wind'—'this is a sacred place' for the people living around the Laguna Superior who 'come to the Barra [de Santa Teresa] to do their rituals.' Continuing that:

> Even though the Zapotecs are in the north, the Ikoots are in the south, it never crossed the mind of the government that people from Álvaro Obregón would unite with people from San Dionisio [del Mar] and San Mateo [del Mar]. Why? Because historically these people have been passive, they are a peaceful people, but the government hit them where it hurts in the Barra [de Santa Teresa]. The [Mexican] government did not expect this conglomeration of forces. Because the opposition was traditionally in the northern part of the *Istmo*; people who are politically aware, violent people, well not violent, but who defended their rights historically and yet they sold their land [to the wind companies]. There was a different way of thinking here [in the south]. First, they went into San Mateo del Mar, they entered over here [pointing to the map]. They went to the assembly; they proposed entering from here to build in Santa Maria and on the Barra de Santa Teresa. San Mateo was the first town that said: 'No!' The assembly said: 'You will not enter.' So their entrance was cut off on that side. Their second option was 'let's enter through Álvaro Obregón.'

While the Barra de Santa Teresa (Barra) is in San Dionisio County, Álvaro Obregón or *Gui'Xhi' Ro'*, which means a small hill in Zapotec, is the town at the entrance of the Barra. In 2007, the Spanish consortium Mareña Renovables began conducting negotiations with politicians while later surveying the area in 2011 to construct 102 wind turbines on the Barra and another thirty on the Pacific Ocean around Santa Maria del Mar. Also included would be the construction of three electric substations; submarine cable (less than one km long); fifty-two kilometers of high voltage transmission lines from the Santa Maria Substation to Ixtepec; and six docking stations to facilitate maritime access, as well as improving old and constructing new roads (IDB 2011). Backed by climate change legislation (Howe et al. 2015), the Mareña Renovables wind park is the largest wind energy project in Latin America at 394 megawatts (MW), and the only wind park in the world proposed on a sand bar (Howe 2014). The project received loans from the United Nations CDM and Inter-American Development Bank (IDB) alongside investors from the Macquarie Group, Mitsubishi, FEMSA and the PGGM Dutch pension fund that all sought to use the proposed 394 MW capacity to power the operations of Coca-Cola, Heineken, Walmart and Bimbo among other companies (Smith 2012; Howe 2014). However, as mentioned earlier, this wind energy project was confronted by the 'Triangle of Resistance'—the towns of San Mateo, San Dionisio and Álvaro Obregón— who joined together to defend the Barra and the sea. After intense fighting, this struggle continues in the minds of locals even with the project officially halted by a court injunction (*amparo*) on 7 December 2012 by Seventh District Federal Court in Salina Cruz (Petersen 2012).

Sustainable development or 'green' projects that grab land, alter environmental relationships and spark conflict, transfer the control of land and/or natural resources to powerful actors typically originating outside the area in the quest for 'green,' sustainable or renewable energy projects (Holmes 2014). These land and resource transfers involve collaboration with authorities at the international, national and local level with these projects utilizing various forms of coercion and/or deception to achieve their desired goals of resource control and concentration (Peluso and Lund 2011). Álvaro Obregón is a case of green grabbing that exhibits the complicated micro-politics of land acquisition, control and conflict that accompanies development projects (Hall et al. 2015; Borras et al. 2012). The wind energy conflict in Álvaro Obregón provides a glimpse into the reality of wind energy development (see also Avila 2018), but also the plight and politics of the semi-subsistence communities facing mega-development projects. It is argued that wind energy is continuing the destructive processes of the industrial economy in culturally and ecologically sensitive areas of the Laguna Superior by directly threatening the biosphere in the region—resulting in further livelihood insecurity, violent social divisions and triggering what amounts to a type of low-intensity civil war in the town.

Based on participant observation along with 123 recorded informal and semi-structured interviews (of which thirty-three were focused on the conflict in Álvaro Obregón), this chapter begins with a brief history of the town that is followed by a section examining the arrival and subsequent triggers of insurrection against the Mareña Renovables wind energy consortium, now called Eólica del Sur. The following section chronicles the uprising against the wind company, battles with police and the takeover of the town hall that resulted in a type of low-intensity civil war between the *cabildo comunitario* (Communitarian Counsel) and the *constitucionalistas* (Constitutionalists). The next section examines the different perspectives and social conflict within the village and how this battle between the Communitarians and the wind company continues. The change introduced by wind energy development is resulting in serious livelihood and cultural changes for the Zapotec, Ikoot and others living from the sea that will intensify pre-existing poverty and dependency on wind farming, and meanwhile slowly encourage cultural transformation to a point of destruction, which is a topic revisited in Chapter 6.

ZAPOTEC STRUGGLE: GUI'XHI' RO'

Álvaro Obregón was established as an agricultural military colony in 1930 by Heliodoro Charis Castro. With 400 troops, Charis sought to impede control exerted by the Oaxacan governor over the *Istmo* by expropriating the estate[1] of Baron Maqueo to create a home and a political power base in what was to be known as Álvaro Obregón or Gui'Xhi' Ro in Zapotec (Monjardin 1993; Smith 2009). Before the Mexican Revolution, the *Istmo* was dominated by ethno-political conflicts between the Reds and the Greens (Smith 2009; Rubin 1997). As a Zapotec teenager in the pre-revolutionary Mexico, Charis began his political career fighting the mestizo Reds—the elites from Oaxaca and Mexico City—by aligning himself with the Greens who were from the *Istmo* and fought for their autonomy and natural resources (Rubin 1997). The *Istmo* was in constant upheaval over land, salt flats and authoritarian impositions from Oaxaca and Mexico City, leading historian Lectica Reina to argue that the Revolution arrived early in the *Istmo* (Smith 2009). Leaving to fight in the Mexican revolution, Charis became an ardent supporter of General Álvaro Obregón, later leading a battalion of Juchitecos under him (Rubin 1997). Obregón became president in 1920 and Charis moved up in rank to general, later returning to the *Istmo* as a popular local figure, establishing a village part of Juchitán County; named after his friend and then Ex-President Álvaro Obregón (Smith 2009). Not long after his triumphal return, General Charis emerged as a political boss of the region (*cacique*), subsequently resolving the conflict between the Reds and the Greens in the 1930s and emerging as popular icon of the *Istmo* (Monjardin 1993; Smith 2009).

General Charis 'established a distinctly local *cacicazgo* around the town of Juchitán de Zaragoza based on traditional qualities of charisma, local knowledge, friendship, kinship, and force' (Smith 2009, 142). Impressed by the (relative) dissolution of the Red and Green political conflict and with state unification in mind, Lázaro Cárdenas's presidency confirmed and promoted Charis's political control not only over

Juchitán, but the *Istmo* region as well as cementing his position as *jefe politico* (political boss) (Smith 2009). Nominally a part of the PRI, Charis acted as the *cacique* of the *Istmo* for twenty-five years, negotiating modernization and the Mexican state's involvement in the region which Jeffery Rubin (1997, 45) said permitted the region a 'domain of sovereignty' (Chatterjee 1993). Charis fought for the social welfare of the region, social development of schools, hospitals and running water, and was a strong defender of Zapotec culture (Rubin 1997; Smith 2009). While his position as *cacique* was not uncontested, Charis held this position of political boss of the region until his death in 1964.

A power vacuum was opened with Charis's death, leading to social conflict and land grabs by opponents and government (see Chapter 1). This was matched by further attempts by political parties to harness his popular mythology (Monjardin 1993). These conflicts and emerging tyranny of the PRI soon gave rise to The Isthmus Coalition of Workers, Peasants and Students (COCEI) in 1973 (Campbell et al. 1993). Throughout the 1970s and early 1980s, the COCEI achieved some popular agrarian victories against private property, terminated the Communal Land Commission, established *ejidos* and defended communal land within the *Istmo* (Rubin 1997). In Álvaro Obregón, Charis's widow and daughter associated with the PRI who claimed his assets and sold some of his collectively used land and salt flats in 1964 to Federico Rasgado, who 'employed armed guards to prevent peasants from fishing and gathering firewood and paid salt workers less than minimum wage, denying them social security and threatening them with violence' (Rubin 1997, 114). This resulted in repeated beatings and later killings in the town, as people rejected this privatization of social property that many felt tarnished the legacy of Charis, who according to local legend, told the town's original inhabitants that 'six hectares and a bullet' was all that was needed to protect Álvaro Obregón (Campbell, in Rubin 1997, 114).

The Leaders of the COCEI, Hector Sanchez, Leopold de Gyves and Daniel Nelio among others would soon join and take on this land struggle with the People of Álvaro Obregón (Rubin 1997). In 1977, the COCEI occupied the headquarters of the Ministry of Agrarian Reform in Mexico City, eventually leading to a presidential decree awarding one thousand hectares to the farmers of Álvaro Obregón, establishing the pro-COCEI Ejido Emiliano Zapata (Campbell 1994). This victory, combined with years of militant struggle against the PRI and Mexican state, would eventually transform Álvaro Obregón into a COCEI stronghold—a place of not only popular support, but also militant armed cadres who would defend the popular interest of the COCEI against other competing paramilitary groups in the region. COCEI success in the *Istmo* came at a high cost of violent repression at the hands of the Federal, state and extra-judicial forces[2] that resulted in frequent beatings, arrests, disappearances and killings of COCEI supporters. Two years after their 1981 electoral victory in Juchitán, the military would invade the city to occupy it in 1983 (Rubin 1997; Campbell et al. 1993). Against the backdrop of this violent and deadly repression, the COCEI began cooperating with the PRI in 1986 and President Carlos Salinas de Gortari in 1989, signing the Pacto *de concertación social* that granted the Juchitán regime, led by Héctor Sánchez López, Federal funds (Rubin 1997). This began COCEI's integrated cooperation with the federal government and who, in the 1990s and 2000s, began permitting the arrival of transnational companies into the region, which would later include transnational wind energy companies (Altamirano-Jiménez 2014).

The 2003 USAID sponsored report, *Wind Energy Resource Atlas of Oaxaca,* designated the *Istmo* as an 'excellent' location for wind energy generation (Elliott et al. 2003). This provoked a wind rush in the northern part of the coastal *Istmo,* which later led to the construction of 1,608 wind turbines with more planned (Rivas 2015). The quantity and strength of the wind in the *Istmo* had wind energy developers attempting to spread wind turbines across the entire region, planning wind parks everywhere, including in and around the Laguna Superior and Inferior (Rivas 2015; Bessi and Navarro 2014). Then, in 2004 in San Dionisio del Mar, without the knowledge of its residents, the *comisariado* (collective land commissioner) Alvaro Sosa accepted a bribe and signed a preliminary contract with Spanish corporation Preneal (Smith 2012). This paved the way for PRI affiliate and mayor of San Dionisio del Mar, Miguel López Castellanos, to allow Mareña Renovables to begin construction on the Barra in 2011 for an alleged payment of between 14–20 million pesos (USD $1–1.5 million) (Smith 2012). Appropriating the name 'sea people'— *Mareños*—into their name,[3] Mareña Renovables, now called Eólica del Sur, tried to enter Álvaro Obregón sparking conflict, violent division and eventually turning the majority of the town against the COCEI and all other political parties.

WIND COMPANY PENETRATION IN ÁLVARO OBREGÓN

Mareña Renovables came to Álvaro Obregón to survey the area in 2011. While there are private homes and land in Álvaro Obregón, the area around the town consists of two *ejidos*: Zapata and Charis communal land and roads that lead to the Barra de Santa Teresa. Social property—ejidos and communal land—appeared in Article 27 of the 1917 Mexican constitution. Ejidos are lands designated for residential and agricultural use, governed by local assemblies made up of recognized community members[4] and could not be sold as private property until the alterations to Article 27 in 1992 (Stephen 2002). The Mexican state, under Article 27, retains 'direct ownership of all natural resources' beneath the surface of *ejidos* (MG 2007, 19), positioning *ejidos* as strategic agrarian concessions made by the Mexican state to pacify revolutionary tensions and promote state unification (Smith 2009; Tutino 2007). The COCEI in Álvaro Obregón used the *ejido* to mobilize and strengthen their political power,[5] establishing Ejido Zapata in 1977. In 1981, Ejido Charis was formed by Charis's daughter and the PRI to counter COCEI's support in Álvaro Obregón, but eventually Ejido Charis became 'COCEI partisans' (Campbell 1994, 276). On the other hand, communal land (*comunales*) is related to pre-colonial land claims and is governed by the community with varying rules and relationships, according to regional customs and practices (Stephen 2002). Despite the sustained efforts of the Mexican government to integrate communal land into the regime of private property (Payan and Correa-Cabrera 2014), communities continue to evade these attempts as Álvaro Obregón demonstrates. The roads to the Barra, the beaches and other lands were considered communal in Álvaro Obregón, which began to change when authorities from San Dionisio leased the land to Mareña Renovables.

Working with the politicians, *comisariado* and Ejidatarios from Zapata and Charis, the wind company began moving into Alvaro Obregon. The politicians[6] and *comisariado* were paid millions of pesos to facilitate the arrival and adoption of the project; meanwhile, each Charis Ejidatario received 40,000 pesos and Zapata Ejidatario received 18,000 pesos (USD 1,038). According to one fisherman, when Mareña Renovables entered the area in 2011 'they just wanted to tour the Barra', offering to pay people 2,000 pesos (USD 115) per day and 100 litres of gasoline as well. At that time, residents claim they were unaware that Mareña 'wanted to kick us out of this town', so the people worked with Mareña without the full knowledge of their intentions. Locals with motor boats started measuring the depth of the Laguna for the wind company, while later, others started digging and measuring the depth of the fresh water under the Barra. '[Mareña Renovables] said to the people of Álvaro Obregón: "We are going to pay you to check if there is fresh water in the Barra"', recounted a fishermen, who continued that, 'we thought they were going to take a hand shovel and dig in the sand to look for fresh water, but by the time we had realized what was happening this company had brought in a generator, hired a tractor and brought in about forty 4x4 vehicles [. . .] and they brought several other machines.' It became undeniably clear that this was about more than just tours and monitoring the Barra.

With the start of construction, the fishermen began witnessing immediate environmental impacts. The same fishermen who helped measure the Laguna explained: They drilled in the Barra, seventy meters deep.[7] From the very first time they began to perform this work the fish began to die. Another fisherman recounts that when Mareña Renovables:

> did testing, they pounded the ground and tons of fish died and went away. [. . . .] Because of the noise and also the oil they were throwing into the sea.
>
> **AD:** *Oil?*
> Oil from the machinery.
>
> **AD:** *Did you see the oil in the sea?*
> Yes.
>
> **AD:** *Could you tell me more or less what you saw?*
> Well at that time tons of fish died—throughout the whole sea as far as that hill—tons of fish died and went away. So when we went to fish, there was nothing to catch. Since they left and there is no more noise, the fish are back, now people are catching fish again.
>
> **AD:** *How long did it take for the fish to come back?*
> Like a year. Maybe they hid, maybe they went away, but tons died.

This account was echoed frequently by older fishermen. Often accompanied by tears and intense emotions, they described the different types of fish, which ones died first, and how millions of fish were scattered across the Laguna as far as the eye could see. Another fisherman equates this phenomenon with an earthquake: 'any kind of sound will scare them away—the fish hear everything.' Fish are vulnerable to habitat fragmentation, noise, vibrations and electro-magnetic interference from off-shore wind energy construction and operation (Tabassum-Abbasi et al. 2014, 282–284), which combined with people dumping oil and possibly other chemicals into the ground.[8] When the construction workers 'changed the oil in the machinery, they just tossed it on the shore. So when it rained, it washed all the cans and oil into the sea. Then you could see all of that oil in the water.' Fishermen witnessed this as they struggled with survey stakes that would cut their fishing nets. The locals rejected the environmental impact assessments put forward by the companies, as they were not consulted and key information was neglected (IDB 2011). As this would be the first wind park built on a sand bar (Howe 2014), information regarding oil, foundation depth and overall impact in this area remains unknown, but as disclosed in interviews, the foundation depth was said to be seventy metres. Confrontations between construction workers and fishermen were the beginning of what would soon turn into an insurrection against the wind company, their political affiliates and gunmen.

The mass-death of fish was an enormous blow to the town, which was compounded by three other factors. First, the town of Álvaro Obregón was never consulted: 'They were buying land, buying roads, I don't know how or who gave them permission, because they did not speak with the community.' Another person contends: 'They entered in a sneaky way. They did not do a citizen consultation. They went directly to the political leaders, the *ejido comisariados* passing out a lot of money—offering millions of pesos.' Mareña Renovables, guided by local politicians, only consulted the *ejidos* and land owners, placing the fate of the town of roughly 3,558 people into the hands of about 350 people (SEDESOL 2010). Likewise, specific information about benefit sharing and environmental impact of the project was deliberately withheld from both the Ejidatarios and residents of Álvaro Obregón (Howe et al. 2015). Large portions of Álvaro Obregón and surrounding populations were not consulted regarding the project development and implementation. Second, there was the unequal pay between the people informed about the project, specifically Ejido Charis and Zapata. For example, the difference between 40,000 and 18,000 pesos between the Ejidatarios in Charis and Zapata, and those who sold their land previously who were not compensated. On the latter, the people in Ejido Charis who sold their land got 20,000 pesos, while twenty-five Zapata Ejidatarios did not receive anything—they were the first to blockade and protest the wind company. Finally, Mareña Renovables placed a chain across the road entrance to the Barra, hired security guards to protect the construction site, and imposed a strict fishing schedule requiring the townspeople to show their identification cards to access their communal land. Reminiscent of the times when Federico Rasgado imposed controls over the land purchased from Charis's daughter, this immediately triggered a popular struggle against the company. The next section will chronicle this before discussing the current perspectives and situation in Álvaro Obregón.

'WE ARE THE SEA': BATTLE FOR THE BARRA AND MUNICIPALITY TAKEOVER

Mareña Renovables blocked the entrance to the Barra and subjected the fishermen to police controls and harassment. The fisherman, 'El Vato', recounts:

> The people wanted to go fish and the state police had a chain there blocking, and they start asking people: 'Where are you going and when are you coming back?' This one fisherman asked them back: 'Who are you to ask these questions? We have been living here a long time.' They replied: 'There is an order here that says you have to tell us when you are going and coming back and that you are not allowed to stay past that departure [time].' So that guy did not say anything else, he just returned back and told people and people were waiting to see what the Mareña Company's next move was.

The 'chain' is frequently recounted in all of its symbolism as the uprising's trigger. Explaining the significance of this imposition, an elder Wild Cat explains: 'Right now we have freedom to go whenever we want [to the Laguna]. We could go at midnight, at sundown or at sunup—anytime we are free, because we do not have a commitment to anybody but ourselves'. Taking great pride in the communal land and the freedom it garnishes, people pray, socialize and fish at all times of the day and night on the Barra. It is the village norm to fish at two, three, four, five or six in the morning, not only in accordance with tide and fish cycles, but also to catch the bus at six in the morning in order to sell fish at the market in Juchitán.

Meanwhile, Mareña Renovables paid workers to clear and improve the communal road[9] to the Barra rented to the company by Ejido Zapata and Charis (so heavy machinery could enter). Shocked that Mareña Renovables went straight for the sea, an Ejidatario remembers:

> That is when the announcement was made to go yell at the *comisariado*: 'Did you sell the sea!?' He said: 'No, no one told me that it was going to be on the sea.' And the people said to him: 'We are sorry Margarito, if the companies take over the sea and prohibit us from fishing we are going to kill you'. He said: 'No, but they did not tell me'.

The wind company and state police moved in and arrested some farmers and fishermen in the vicinity, resulting in a town-wide announcement on the loudspeaker. A woman recounts: 'The people that had their relatives working out on the field, they need your help; they have just been kidnapped!' They [the people] grabbed their machetes, sticks and their slingshots' and eventually 'when that happened the people detained the two *comisariados* and their children and the people said [to the police]: 'If you release the people that you have, we will release the comisariados. If you don't, we will kill the *comisariados*.' Conflict erupted. People went to the Barra and began attacking security and the equipment of the wind company, burning some of their vehicles. The first skirmishes with state police began in late January, before the Battle of the Barra on 2 February 2013. The barricade was formed in General Charis's abandoned house (see figure 4.1), and 'people said they would all be willing to die before allowing the company to go past the barricade', recounts the captain of the communitarian police. Summarizing earlier police interventions, he explains: 'the company tried to enter with fifteen police patrols full of state police. They couldn't and they tried again, this time with twenty-five police patrols. Then they tried again with more than fifty patrols and they couldn't'.

The state police tried again on 2 February at eight in the morning.[10] Attempting to flank the barricade, roughly 500 state police came in from the north near Ejido Zapata, while another regiment came from the east near the cemetery and the Laguna. Learning from the previous battles, the police entered the town on foot shooting their guns into the ground and air. Trying to find contacts in Oaxaca City to avoid a confrontation, the town negotiator contends, 'it was impossible to sustain peace, a confrontation had to take place'. Driven to fight based on livelihood needs—'if you take the sea, then how are people going to live?'—and a generational commitment to the land and sea: 'I am fighting for my people, for our sea, our land, our children, grandchildren and future generations. [. . .] The people do not realize they are accepting money from the wind energy company in exchange for losing the inheritance they would leave to their children and grandchildren.' The people of Álvaro Obregón felt their livelihood, culture and by extension generational lineage was under attack by the arrival of the wind energy project, and now faced a small, well-equipped army of public and private security forces. The captain of the *policía comunitaria* explained that this was to defend Genera Charis' legacy. Recounting what General Charis told his grandfather right before he died:

Figure 4.1. The Communitarian Headquarters, 2013: Veredas Autónomas

Several *compañerxs* shed their blood so I could have this land and so I leave this colony for you, for the people of Álvaro Obregón. So if some other motherfucker comes trying to take away this colony, you have to fight for this land, because I looked for good land for you, for all of our people.

When the police moved into town the people realized 'they were surrounded, but people were already prepared with their slingshots', said El Vato, while The Dragon, another Communitarian, remembers:

[W]hen I got back to the barricade the people had a bunch of fireworks and I said: *compañarxs*, now is the time to start the war. And they said, 'Where are we going to start?' And I said, 'In my opinion we should start against the state police at Ejido Zapata [. . .].' And we called people: 'Who is going to go?' And everyone said: 'I will!' Not ten, twenty, or thirty people, everyone said they were going to war. So we went to fight about 500 metres from the barricade, and that is where we found and fought them. I set off a shitload of fireworks at those idiots (*pendejos*)! And my *compañarxs* were throwing rocks and I was shooting fireworks and we chased them away from the 500 metres where we started, we defeated them, pushed them away—away from there. And we came back and took the other road to the cemetery where the other contingent was, they were all on the beach, but they wanted to come back to the cemetery, but we had another barricade in the cemetery.

The Turtle, another member of the Communitarian Police, recounts a similar message:

We have to fucking fight—fight time. We cracked some of the state police skulls. They got a good beating, like this—bam! And all the stones thrown and shot, we ran out of stones on the beach and then we started using the shells with sharp edges; we used everything against them. I grabbed the shell, even with the sand in it and I would shoot it at them—boom! I would grab the sharp edge and it would cut me, but I did not care. I was just making mayhem—Boom! Boom! Boom! There were a lot of people fighting, there were even kids fighting—tons of kids. That day it was a tough fight. [. . . .] Those guys [police] were bleeding and on this side, these guys also went out toward Xadaní—that is the way those teenagers went, they did not go through the town directly. They went around the town, but the people in Xadaní were also waiting for them, and they saw that and they ran around, otherwise they would have been killed by the people from Xadaní [. . .]. So from then on, we stayed in the barricade, we were there for a year and we stopped the elections—boom, boom—we do not want anyone here, which is what we are saying, even now. We are still fighting.

The people of Álvaro Obregón organized and rose up to defend the sea, which the Tank said was won by 'using fishermen wisdom, we started to move just like when we are fishing and started to surround the police. It's important the fisherman wisdom, and with slingshots we made them run because they knew that we had them.' The fisherman wisdom is culturally rooted knowledge serving to influence their approach to battling police. This victory led to the formation of the *policía comunitaria* (communitarian police, see figure 4.2) and the *cabildo comunitario* (community counsel), known as the Communitarians. The Communitarians realized that the next step in their struggle, if the Barra and the sea were to remain safe from development, must be to expand the struggle to include *all* political parties who collaborated together to allow the wind company to lay siege to the town.

Despite the popular resistance displayed during the Battle for the Barra and the issuing of a court injunction, the political parties backed by Mareña Renovables began developing a counter-force to combat the Communitarians. This struggle over the Barra became fused into past feuds such as allowing workers access to one of the largest salt refineries in the region, which was also located at the entrance to the Barra. The ejidos members and their relatives were favoured by the salt refinery owner, becoming another issue for the Communitarians that overlapped almost seamlessly with the civil conflict around wind energy development on the Barra.

Oaxaca is unique compared to other Mexican states in its recognition of Indigenous customary law—*usos y costumbres*—and semi-autonomy. Relating to pre-colonial recognition that Indigenous people had localized forms of governance and justice, *usos y costumbres* reemerges as a form of Indigenous customary law and governance that literally translates as 'practices and customs'. This recognition of Indigenous rights and semi-autonomy in Mexico began in the mid-1980s in the face of structural adjustment and democratic reforms, which began affirming governmental decentralization and what Hale (2002) had called 'neoliberal

Figure 4.2. Communitarian Police leaving for patrol

multiculturalism' (Ulloa 2013/2005). Despite the statist motives for these changes, they were tied with real aspirations for self-determination and autonomy to ensure the survival of the Indigenous populations. This aspiration was advanced in the 1996 San Andrés Accords negotiations between the Zapatistas and Mexican government. Despite the Mexican senate approving a weaker version of these accords, many of the core ideas had been established in Oaxaca previously (Stephen 2005). In the early 1990s, there were two modifications made to the Oaxacan constitution: (1) Article 16 that recognized the sixteen different ethnic groups in the state; and (2) Article 25 establishing 'respect for the traditions and democratic practices of indigenous communities' (Hernández Díaz 2007; Stephen 2013, 69). Then on 21 March 1994—Benito Juárez's birthday—the state government announced the *Nuevo Acuerdo para los Pueblos Indíenas* (New Agreement for Indigenous Peoples), which recognized *usos y costumbres* and advanced the rights of Indigenous people in three areas: the administration of justice, resolution of agrarian conflict and respect for customs and traditions (Stephen 2013). This continuum of Indigenous struggle and legal context in Oaxaca undoubtedly influenced the Communitarians' next move.

On 8 December 2013 (APIIDTT 2014), the Communitarians peacefully seized the Álvaro Obregón town hall. Deciding that they are not going to let 'the politicians give orders here anymore—it is time for us to give the orders', over 1,000 people marched from the General Charis barricade to the town hall. '[T]he Polícia Comunitaria led the way followed by the [Communitarian] mayor with his council', recounts El Vato, who then explains how:

> we walked over here and we stopped before we got to the town hall [in the town square]. There were police guarding the town hall and we talked to them and we said: 'You know what? We are going to be taking on this work guarding this building now. We think that your year of service has now ended. You should just step aside and let us takeover.' And they said: 'That is fine, we are leaving.' They grabbed their stuff and they left and we stayed here. [We asked each other:] 'Now what do we do? What about the [police] patrol vehicles? Let's go talk to the [former] mayor Ricardo and ask him to turn the vehicles over to us officially and voluntarily, and also the ambulance and everything else. Okay, let's go.' Some of the folks went in like a delegation and when they talked to him, right then he created a document and signed it, turning over the vehicles voluntarily. So the new [Communitarian] mayor came in and started to work.

The Communitarians took over the town hall establishing the *usos y costumbres* system of governance historically designed around communal relations while articulating a form of *comunalidad* (Luna 2010; Manzo 2011). This governance system is designed around *civil cargos* (community public duty), *mayordomías* (religious cargos/communal festivals), *tequio* (communal work obligations) and cooperación (communal pooling for collective maintenance/projects) (Stephen 2005; Gross 2015). Among one of the civil cargos is the *topilillo*, or policeman, who serves a similar function as the *polícia comunitaria* following the local tradition of no pay for their communal services, even if the village is divided over political rule.[11] Since

Oaxaca state recognizes *usos y costumbres* the town hall receives state funds which places a higher value on what faction legitimately holds the town hall, a dispute tied up with their legal battle of Indigenous recognition that is currently before the Inter-American Commission on Human Rights.[12]

The former municipal leaders were associated with the Juchitán government (held by the PRD-COCEI), political elites and backed by Mareña Renovables/Eólica del Sur. They were displaced from the town hall, subsequent to which they formed the Constitutionalists—commonly known as *Los Contras* (the enemy/cons). In short, the Contras are the politicians who have created a counter-town hall and police force with backing from Juchitán. They recruited people from the town and barricade with food baskets, money and jobs with the intention to re-establish the political process—voting—according to the Mexican constitution. Hence the name: Constitutionalists. The Communitarians refused any elections as they see the political processes as completely rotten and corrupt, with politicians filling their pockets with money, buying votes and systematically impoverishing the town; something that they had experienced for many years before the recent battle for the Barra. Most importantly, the political process and politicians were surrogates who legitimized the building of the wind energy project on the Barra, resulting in and by extension significantly degrading the mammal and marine habitats. This has resulted in something akin to a low-intensity civil war—filling the village with tension—with two factions patrolling the town with their shotguns and M-1 carbine rifles while screaming threats and fighting using fists, sticks, machetes, slingshots and on occasion firearms. The lower manifestation of this tension, a woman explains, was a neighbour 'saying nasty things to us, saying he was going to come into our house and every time we walked by he yelled and threw stones at us.' Such occasional high-intensity confrontations usually become more common around election time.

The Mayor of Juchitán Saúl Vicente Vázquez announced elections to be held in Álvaro Obregón on 2 March 2014.[13] The Communitarians boycotted these elections and began patrols at six in the morning, effectively preventing the state police and representatives of the National Electoral Institute (INE) to gain access to the town. This resulted in some high-speed chases, arguments and slingshot attacks to keep them from entering the town, lasting until about eleven in the morning. Then the communitarian police regrouped and celebrated with a big party at the town hall complete with piles of rocks, iconic Zapotec women sitting in traditional dress with enormous sticks and food, enjoying themselves as they prepared for battle. Meanwhile, members of all the political parties from Juchitán were holding elections in the house of the local political candidate. Around two in the afternoon, a group of Contras and gunmen from Juchitán emerged from the elections house, and began attacking the Communitarian town hall with bullets, stones and sticks (APIIDTT 2014). This erupted into an intense battle, complete with rocks falling like rain, western style-shoot outs, and barricade fighting in the streets. An anarchist in the town fighting alongside the Communitarians recounts:

I was running down that street going to the entrance of the town. The entire way along there were people yelling, rocks falling, glass breaking, a lot of screaming—it was a very loud situation. When I ran to the front line I suddenly heard—'POW! POW! POW!' I was like: 'FUCK!?' I turned my head up and there was a *compa* on the roof of the town hall with a 22 [rifle]—'click, click. POW! Click, click. POW!'—and I turned to my slingshot and I was like, 'fuck' and kept running. So when I was heading to the frontline the whole line of *compas* broke, they ran into the walls because there was shooting from the other side also—'POW! POW! POW! POW!' I was like: 'FUCK!' When I arrived I realized that my slingshot was not powerful enough because I was shooting them and I hit them, but they [the Communitarians] shot their slingshots, the people would fall. So I abandoned my slingshot and started picking up rocks with my hand and fighting.

There were many groups of people in the blocks surrounding the town hall fighting the Contra Attack. In a group of fifteen people, the Turtle explains how 'it was raining stones!' Continuing that the Contras 'would have killed us if we had been too stupid' and how:

Contras had some guys who were shooting [stones] at us from above and we moved as fast as we could to get back to where everybody else was and when they saw us coming they came out with sticks, stones and slingshots and everything—Boom! Boom! [. . . .] Some of our *compañarxs* were hurt. I was also hurt because I was hit by a piece of cinderblock right about there [point to the back of his head] and I grabbed another stone and bent over and another piece of concrete hit me right in between the shoulder blades and I was falling down—for real. So I stood up and kept shooting stones. So we were all there and that is when this guy the Vato covered me. So I could get into a house and I broke into the house. They had just finished that house and we broke in, [. . .] we went inside and they kept shooting stones at us, but since the window was wide we shot from there and they could not come any closer. [laughter] We were in the window—BANG! BANG!—we were dropping them and making them look like idiots, that is when we fucked them over.

The Communitarians held the town hall,[14] but later that evening Contra forces broke into the house of the Communitarian police captains, pouring gasoline everywhere to burn it down (APIIDTT 2014). This resulted in another intense chase and battle, but after which the Communitarians were victorious. While officially, the Mareña Renovables/Eólica del Sur project was halted and moved to a new location north of Juchitán near La Ventosa, ushering in a phase co-existence in Álvaro Obregón between the Contras and Communitarians, until the next election called for 7 June 2015.

THE CONSTITUTIONALISTS—THE CONTRA

'The Contras are the politicians', explains a Communitarian. When it is time for elections in Álvaro Obregón the politicians go door-to-door, holding rallies and distributing food baskets as well as pay 'people two-thousand, three-thousand pesos for their vote'—while voters hand over their voter identification and information. Rooted in family and patronage networks, the politicians create a political cadre by providing promises and 'support'[15] to the town. When Mareña Renovables, among other wind companies, came to the *Istmo*, this appeared as a lucrative opportunity for politicians. 'Now that he has nothing left to grab', says Manpache discussing a member of the COCEI, 'he has come here to take our lands from us and sell it'. Mareña Renovables came with money and the politicians were quick to negotiate and become their proxy force assuming the loyalty of Álvaro Obregón. The politicians with funds from Mareña Renovables were supposed to stabilize and manage the political terrain for wind turbine development. Impersonating the wind company, Pingülino sees the relationship between the Mexican government and the wind company: 'Hey, I already paid you to put your people in line; it's your people'. The political parties, notably the COCEI because of their history in the town, acted as intermediaries, managing a type of 'decentralized household rule' and, later, with the constitutionals, a proxy force to counter and domesticate local resistance fighting the wind energy project (Owens 2015, 8).

In the context of peacetimes, the 'counter organization', expounded by Brigadier General Kitson (2010, 79) in *Low Intensity Operations*,[16] is a 'method by which the government can build up its control of the population and frustrate the enemy's efforts at doing so'. This can be done with arms and civil assistance—social development—to stifle opposition groups competing for legitimacy in an area.[17] This was the intention of the Contra police force, backed by the state government and transnational wind companies, who were able to recruit people from the 2 February barricade. León, stated: 'there was a kid who was part of the Communitarian Police and they [the Contras] bought him off with a pistol and 1,000 pesos a week salary and he went with them. That kid betrayed the movement, so that is how they know how many weapons we have here'. In this context of material poverty, popular media saturated with narratives and images of luxury, comfort and depictions of extreme violence in television, a gun and this type of money creates new opportunities to fulfill material desires.

Constitutionalists, backed by the Juchitán government, are comprised of roughly four overlapping types of Contras that make up their loose-knit counter-organization. The first believes in modernization, hoping that the wind project will fulfill their dreams of progress with jobs, social development and prosperity. Second are the fishermen and farmers who want jobs to feed their family and improve their lives and are willing to work with the Contra police, as long as they are not forced to shoot their neighbours. Third, political party militants are the people who are loyal to the party ideology and embrace the actions and imperatives of the group. Fourth, the mercenaries/gunmen hired to make and, at times, participate in this counter-organization and thus transform it into a trained police or paramilitary force. The majority of Contras, however, fall into the second and first categories. The Constitutionalists in the first category, typically younger, are bored of the traditional lifestyle, arguing:

(1) the fish population is declining and (2) that the majority of the income from the town comes from seasonal migrant labor within and outside the region (Huatulco, Cabo San Lucus, Loretta, Monterrey and the United States), justifying wind park construction and land change.[18] This perspective, the Communitarians tell me, is a product of education and school attendance becoming mandatory in the area, holding the school responsible for cultural change and even disintegration.[19] An experience shared, albeit in a different intensity, with Native North Americans boarding school programmes (Woolford 2014).

Regarding the second category, in an interview with someone who had a one year contract with the Contra police states:

> I was working for them because I have a family to feed and they were paying me. [. . .] I have worked with them because they paid me to help maintain my family, but the reality here is that everyone from here is like brothers and sisters and what is the point if I run into somebody and we fight? The politicians will just be standing off to the side and watching us fight—this makes no sense.

Many people associated with the Contra faction are against the wind company building on the Barra but do not like the municipality split, favouring voting—or have vendettas against individuals or families apart of the *cabildo comunitario*. The Contras often emphasize the combative attitude of the Communitarian police and their propensity to drink and smoke marijuana. The third faction of political party militants are people who prefer to work with the Contras because of access to welfare benefits, food handouts, payments, family ties plus a generational ideological belief in the COCEI or PRI. Finally, the *Istmo* in general and the Seventh Section neighborhood of Juchitán are known for having gunmen, families, elites and politicians that organized hit squads and paramilitary around Oaxaca and other parts of Mexico. In Álvaro Obregón it has been reported that there are extra-judicial gunmen operating in the service of the political parties, from outside and inside the town, indirectly supported by wind companies—it is left to local politicians to manage, stabilize and secure the area for wind park construction. It was also rumoured that a policing and counterinsurgency expert was hired by the Juchitán administration to train the Constitutionalist police to enhance their police professionalism. Explaining this activity, the Flower, another Communitarian, says, 'the Contras from other places are setting fire to people's houses. They are burning [fishing] nets of poor fishermen because we don't have money and they do.' Nevertheless, a good portion of the people I talked to were confident that if Mareña Renovables/Eólica del Sur tried to enter the town again, the majority of the Contras would rise up with the rest of the town.

THE COMMUNITARIAN STRUGGLE CONTINUES

The impact of wind energy in Álvaro Obregón has been significant. Wind energy is a new conflict in itself that ignites and widens old fissures over land going back to divisions between the COCEI and PRI, if not to the Revolution itself. The current manifestation of this generational struggle is not only to protect the land and sea, but also for the next generations to live free and dignified Istmeño cultural lives. Before going to pick mangoes, Big Bear says:

> Compañarx, to tell you the truth, we are watching over the sea and territory of Álvaro Obregón—look at this beautiful wind, lots of birds singing all the time, and the mangos. See we are going to lose these animals and we do not know what else, why are they [the wind companies] going to do this? Over there [in the Cities], you have to pay for light, water, you have to pay for everything and you are going to have to work like a slave to earn 2,000 pesos and you pay 2,000 for rent for one month—how are you going to eat? Then 600 pesos for water, and then the next week you have to pay 600 pesos for light, and then you will pay 800 pesos for eating for one or two weeks—I have already been there and I have seen it.

Describing modernized life, Big Bear recognizes a style of slavery and an inherent separation from the natural environment—a separation that naturalizes industrial corridors and degraded landscapes. This fight against wind energy is not only about the preservation of the beauty and cultural life attached to their lands, but also a desire to fend off a type of modernization placing them in a *greater* position of dependence and subjugation. The Communitarians are not against development or progress, but any type of development that will degrade the sea, threaten their livelihoods, and place their future existence in the hands of transnational wind companies or, in their own words, 'foreigners'. Furthermore, the Communitarians are not against wind energy per se, but the location, quantity and how the energy will be used. While Álvaro Obregón is divided, the Communitarians and even a portion of the Contras do agree to protect the sea, develop on terms that respect the land and the entire community; something politicians and influxes of money seem to disrupt and prevent from becoming realized. 'Government welfare programmes are like fishing: they put out bait for people to grab thinking it's food, but it is always to fuck them over', explains The Lizard, who concludes his point by saying: 'The government will never lose when they give money to you.'

The Cabildo Comunitario also has divergence within its group; there are three main overlapping factions. First, people that want to protect the land and sea and rid the Contras from the village—'We need to break each one down until they are destroyed', say the Dragon, '[s]o the town can be completely free of politicians, foreigners and the big companies that do not want to pay and come in right.' Second, Communitarians who want to reform this division between Contras and Communitarians—destroying the Contras 'is not that easy because we are all the same family and we are from the same town.' Third is the never-voiced position of people who fight to get more money, social development and a better contract with the wind company and/or politicians. There are two noticeable 'neutral' positions articulated in the interviews within the town. The first wants the land and sea to remain protected, but feel they are stuck between two gangs—Constitutionalists and

Communitarians—who are fighting for control over the town. This emerges from the general combativeness of the *polícia comunitaria*, mixed with drinking and preexisting personal conflicts. The second neutral position emerging from interviews favours the wind project, but is critical of the benefit-sharing, social development and political corruption in the town. This leads people of this persuasion to criticize the Communitarians for not voting or participating in formal elections and the Contras as 'selfish' and 'corrupt' politicians. Considering their critique of political corruption, the latter criticism is contradictory and resonates with the Communitarian critique of politics to establish direct democracy and community control based on *usos y costumbres*. Notably, a *muxe* related to a woman articulating the first neutral position, asserts: 'I do not care what the others say [about the Communitarians], because once some of the people from the other side [Contras] beat me up and they did not do anything about it, and this is why I do not pay attention to them [the Contras].' This reflects a type of violence that goes unacknowledged and unreconciled, and influences how the residents pick sides within the conflict. Eventually, taking on the quantitative logic of democracy, I began asking people in the town who holds the majority of support—Contras or Communitarians? The responses asserted unanimously that it was the Communitarians, with one woman going so far as to claim that they have reduced crime and made the town safer since taking control of the town.

This woman's claim that the town is safer resonates with the findings of Sierra (2005) and David Gomez Vazquez (2017), who sees the *polícia comunitaria* in Guerrero as creating safer spaces and realms of participation for women. This is not to say that problematic patriarchal relationships still do not exist in the town, because they do and are linked to the cultural traditions of the past (Ruiz Campbell 1993; see also Gross, 2015), the strong presence of the Catholic and evangelical churches in the town as well as small town gossip that is exaggerated by the conflictive divide over wind energy and political authority in the town. While there is a high and fluctuating rate of female participation in the assemblies according to my observations, the only women who actively did patrols with the *polícia comunitaria* were women supporters from the outside. That said, as in most life instances, the women are foundational players not only in providing food for the people doing *civil cargo* services, but they assume strong combative positions to defend the Barra and the town hall as well. Álvaro Obregón's path towards autonomy, self-reflection and a high rate of diverse supporters of different radical political persuasions from the outside coming in solidarity, will not only set examples, with women interested in the *polícia comunitaria*, but also create reflections and conversations regarding the challenging patriarchal relationships that have the potential to create personal and collective action. Nonetheless, the high level of support and popularity for the *cabildo comunitario* and *polícia comunitaria* undoubtedly comes from its defense of the land and sea, which is reaffirmed by the reality and experience of people living near or surrounded by over 1,608 wind turbines in the Northern part of the region (see Chapter 2).

The people of Álvaro Obregón are all too aware of what Big Bear reminds me: 'We are poor, but you do not die of hunger here', which is precisely what the wind company threatens. The construction of 102 wind turbines on the Barra and the correlating marine and ecological degradations threaten the access and availability of food from the sea, but more so the quality of food, risking *greater* dependence on imported factory fish or canned food. This increasing dependence is being experienced to various degrees in towns engulfed by wind turbines in the northern *Istmo* like La Ventosa (see Chapter 2). Many of the people in Álvaro Obregón know the vital importance of the Barra to their existence and culture. They attempt to communicate this situation in economic terms to foreigners, referring to the Laguna as their bank account (Howe et al. 2015)—'I tell them I have a bank, I get my money form there without a card or anything when I want to get food for my family.' This technomorphism, equating the sea with a bank account, unwittingly plays into the market-oriented trap of payment for ecosystem (PES) services that seeks to quantify and price the natural environment to integrate nature into the financial grid of legibility, so it can be bought, sold and traded (see Sullivan 2009; 2010; 2017a; Brock 2018). Demonstrated here is the economic hegemony that influences people to think of the natural world in economic terms, altering people's relationship and engagement with their environment, transforming relationships of reciprocity to exploitation and degradation in exchange for the material accumulation of wealth, and often survival under the terms set by capitalism.

Despite the previous failures, the struggle continues against both wind companies and political parties. While the Mareña Renovables wind project was officially halted in December 2012, it was rebranded as Eólica del Sur and moved to a location north of Juchitán and approved by a Free, Prior and Informed Consent (FPIC) consultation in June 2015 that will be examined in the next chapter. This wind project was later temporarily halted in October after a court injunction (*amparo*) that became indefinite in December 2015, prohibiting wind turbine development in Juchitán County (Manzo 2014; Sin-embargo 2015). Unfortunately, the Barra is not in Juchitán County. In the spring of 2015 the government was still trying to build a road from Salina Cruz through San Pedro Huilotepec, past the *ejidos* and out to the Barra. Questioning representatives from the Inter-American Development Bank consultation, they denied any attempts to continue wind park or road construction. Nevertheless, the road construction continues. '[T]hey started a month ago on 8 April [. . .], we went to a plot of land and we saw it,' says an Ejidatario who continues that:

> 'they are clearing and bringing a lot of machinery and I work on that side [of the ejido]. There are also three police patrol vehicles protecting the area and they are not from here and they are well-armed—they are state police and they do not let anyone enter there.

The wind company, Big Bear explains, says they 'are going to return in 2016 and we are waiting for them and we are going to fight them again.' This indicates that the struggle for the Barra is not over, as Ejido Zapata and Charis had accepted the money and Mareña Renovables/Eólica del Sur now has a thirty-year contract with three automatic twenty-five year renewals. An Ejidatario asks: 'The farmers that received the money should give it back to the wind energy company so that they do not come. Is that possible?'

Road construction happens alongside aggression on the political front in Álvaro Obregón. The Mexican government announced elections for 7 June 2015, which was met with a call for an election boycott by the Section 22 teachers union with the support of striking workers, Zapatistas, anarchists and other municipalities and communities, like Álvaro Obregón continuing to fight for its autonomy. Responding to economic restructuring and dirty-war style state violence, daily demonstrations and blockades took place against gas stations, highways and government buildings, among other actions (Matias 2015). This put the Mexican government on high alert, increasing the military, police and paramilitary[20] presence around Juchitán, escalating tensions and in Álvaro Obregón where 'there will not be elections [. . .] whatever the cost', explained the Rana, who continues, 'threats from the electoral authority through the radio and in the newspaper saying that on the seventh the Mexican Army will be deployed to support the elections.' It was around this time that a person overseeing my research in Mexico received information that I was being targeted for an 'assault and possibly fatal "accident"', which was confirmed by people in the community.[21] This was also matched by rumours spreading from the Salina Cruz court house that I was an undercover policeman, which coincided with intimidations and attacks by the Contra police. This quickly led to the collective decision that I would be more helpful to their struggle outside the country, resulting with the termination of my fieldwork on 19 May 2015. Subsequently another civil conflict erupted on 7 June, not with a military siege,[22] but with gunmen from Juchitán who were targeting outside supporters that I would suspect was to minimize blood feuds between families in Álvaro Obregón. This, I am told, resulted in a day-long battle more intense than 2 March, when a total of ten people were shot, six of whom from the Communitarians and three of whom were in critical condition. Fortunately, there were no fatalities.

Álvaro Obregón continues its valiant struggle and, from what I am told recently by a friend in town, after the December 2015 injunction against Eólica del Sur there was an increase of extra-judicial repression on land and even fishermen being robbed in the Laguna. The 2015 court injunction continues to be contested, while the imposition of political parties in Álvaro Obregon has caused an escalation of violence. On 14 May 2016 during a political campaign rally by candidate Gloria-Sanchez (PAN-PRD-COCEI) in Álvaro Obregón, conflict broke out between the PRI, the Constitutionalists and Communitarians, leaving one Constitutionalist police captain dead and more on both sides wounded (APIIDTT 2016; Gobierno 2016). This provocation by the political parties to impose elections in Álvaro Obregón has been condemned by the Zapatistas and the National Indigenous Congress (CNI) (Imparcial 2016). Recently, on 22 July 2018, Rolando Crispín López, member of the APIIDTT, was shot in the back multiple times and killed, in the process wounding an eight-year-old girl in the street (APIIDTT 2018). The struggle against wind energy and for political autonomy in Álvaro Obregón continues with increasing violence, death and uncertainty (see figure 4.3).

Figure 4.3. Town Square Mural: 'Freedom is not conquered on your knees, but standing on your feet. Giving back hit by hit, inflicting wound by wound, death by death, humiliation by humiliation, punishment by punishment. Let the blood flow in streams because that is the price of freedom.'

CONCLUSION

Wind energy, driven by climate change mitigation, is instigating and reigniting old and new conflicts further stressing and degrading the natural environment and many facets of social life. The process of these developmental impositions is complex. Nevertheless, it is clear that development projects—'green' or otherwise—will have substantial impacts on Indigenous populations living from the land and sea. In Álvaro Obregón this caused deep social divisions, escalating to a state akin to a low-intensity civil war, transforming the town into a battleground. The Mareña Renovables project emerges from global climate change concerns that have now made it politically feasible to capture natural resources such as wind for profit maximization while aiding the industrial expansion of supermarket chains, processed food corporations and soft drink manufacturers. The immediate complications arise with notions of property and ownership that allowed Mareña Renovables to seize the land, aided by private security and state police. This was compounded by political corruption that facilitated inequality (or embezzlement), translating into local government, landowners and Ejidatarios, full of hopes of prosperity promised by Mareña, selling-out the collective future of their own town. Once the implications of these choices became undeniably clear, popular insurrection for the land, sea and dignity was put into action.

Mareña Renovables undercut the local people, negotiating with regional politicians, ultimately disrespecting the town by providing incomplete information and negligible monetary compensation—nowhere near the cost and hardships experienced by individuals fighting to maintain their subsistence practices. The Mexican government facilitated and fuelled this conflict. The neoliberal legislation of the past thirty years, now appearing in the guise of climate change legislation, allowed the entry of transnational wind energy companies to achieve their original goals at the cost of many. Wind energy development was managed by proxy collaborators, in this case the politicians, with the Mexican political system legitimizing this acquisition while remaining a silent accomplice to the violence perpetrated by state and extra-judicial forces. Proxy and household governance techniques resulted in violent conflict with minority landholders and politicians controlling the future of the people in the area. The arrival of this wind project has divided *Álvaro Obregón* and incited the local population against each other as they struggle for survival, resources and dignity whether as Communitarians or as Constitutionalist police. Meanwhile, the politicians and their managers, inculcated into the logic of capitalism, sell the people's land from beneath their feet. The Communitarian elder Wild Cat explains: 'When we take their money, in return we are giving them our hands so they can tie them.'

This sentiment is shared by another friend who tells me: If you go to a ranch to make chicken stew:

> the farmer throws some grains of corn on the ground and the chickens come to eat the corn and so he chooses the best chicken and grabs it. The assistance that the wind companies are giving is like that. Once they have the people in their hands, they grab them. So once the wind energy companies have dominated Juchitán, they cause our worldview to disappear.

Green grabbing in the *Istmo* and possibly elsewhere is like grabbing a chicken. The money and social benefits creates divisions, not only allowing space for transnational, federal and local authorities to intervene into people's environments, but combined with poverty will create circumstances that render people as available reserves to arm and transform them into police, paramilitaries or gunmen. The end result for sensitive semi-subsistence cultures is that their fishing grounds, cattle grazing and religious sites will have wind turbines built on them, altering these lands beyond recognition. Land grabs such as this directly affect the ability of Indigenous populations to subsist, which has direct cultural implications for accelerating cultural change in a particular industrialist and capitalist direction, threatening to eliminate past traditions, ontologies and relationships to the land.

Therefore, the wind energy project will enable some people to become wealthy in the area, but force others into a variety of choices: (1) starve; (2) take up jobs where they are available in the region; and (3) migrate to find work in tourist, industrial zones, and security forces or with transnational criminal organizations in Mexico, the United States or elsewhere. Wind energy development intensifies the pre-existing political corruption, poverty and environmental degradation in the area, risking poverty to reach unprecedented levels—collectively threatening the lives of people living from the land and sea. Said differently, wind energy development on the Barra has the potential to exaggerate the destruction of semi-subsistent Zapotecs and Ikoots in the region surrounding Álvaro Obregón, further pushing them into assimilation and into work-related out-migration. If mitigating climate change and socio-ecological diversity is genuinely a priority, industrial development—green or otherwise—needs to be completely re-conceptualized in order to work harmoniously with the people, their lands and creating spaces and resources to strengthening developmental alternatives and commitments to live *with* the cycles of the environment, not *on* top of it, subjugating and destroying environments and the peoples who live from and connect to them. The present trajectory of techno-capitalist progress is heading in the wrong direction, which is only worsening with green economic initiatives branded as 'solutions' to environmental and climate crises.

NOTES

1. *Latifundio*.
2. Committee for the Defense of the Rights of People of Juchitán.
3. This is regarded by locals as an insult and a discursive attack waged against them by the company.
4. That was traditionally all male.
5. Article 27 allows Ejidatarios to have a rifle for hunting and a shotgun to defend the land, providing political parties with armed supporters.
6. Confirming whether this was individually or collectively is difficult from the interviews, but according to one interview, specific COCEI politicians were paid anywhere from fifteen to 125 million pesos.
7. Costal Ecologist Patricia Mora also believes this. Fisherman Interview, 21 January 2015.
8. Constructing wind turbines on sand, vegetation and fresh water, requires solidifying agents in the concrete. A farmer in outside Juchitán described how: 'they brought some fluids and they poured them into the water and I do not know what happened, but after that the water stopped.'
9. This work was organized by the COCEI and workers had to hand over their 'voter IDs, everything', but they were paid very well. One worker got paid 1,200 pesos for two days work.
10. See visuals: https://www.youtube.com/watch?v=IyETkKlJ54w and https://www.youtube.com/watch?v=oFszED7ECs8.
11. There are three different types of community armed groups, from the lightly to heavily armed: (1) *topilillo* (police man), (2) *policía comunitaria* (communitarian police organization) and (3) *autodefensa* (self-defense groups), which are associated with higher calibres armaments as exemplified by groups in the Mexican state of Michoacán. See John Gledhill (2015) for more on *autodefensas*.

12. This claim was filed with the Independent Investigation and Inquiry Mechanism (MICI) of the Inter-American Development Bank.

13. The film, *Istmeño, viento de rebeldía: un documental de resistencia al despojo* by Aléssi dell'Umbria is comprehensive, and briefly documents this battle at the end of the film.

14. For more details and accounts of skirmishes go to the Assembly of Indigenous Peoples of the Isthmus of Tehuantepec in Defense of Land and Territory (APIIDTT) blog entries from 2012–2014.

15. 'Support' is an indirect way to say 'money' in the region.

16. Undoubtedly counterinsurgency has earlier precedents in Mexican history than the late 1960s.

17. Kilcullen (2006, 33) calls this 'armed social work' in the context of Iraq.

18. Friends tell me this is a scripted line recited by the Contras.

19. For a conversation on cultural change through schooling in Oaxaca see Meyer and Alvarado (2010).

20. There was a rise of unmarked trucks with military and/or police personnel.

21. Most of the people I was working with have been targeted by gunmen, state police or the military. Likewise, people resisting these projects have friends and family associated with the Contra, gunmen families and political parties, which allows for privileged information.

22. See Dunlap (2015a), which expressed this concern. However, a friend tells me that after the battle, the Navy was on the outskirts of town and came to check on the town, telling people to call them when there were corpses to pick up.

5

The Theatrics and Violence of Consultations: The Free, Prior and Informed Consent (FPIC) Consultation in Juchitán

> [I]n the consultation there has been several people who say 'Thank you COCEI, thank you city hall, thank you mayor and his administration. We are here at the consultation to give applause to the mayor and his administration, because it is thanks to them that the consultation arrives.' I have been really quiet; I only speak when I know I can hit them all, so that day after everyone talked I took the microphone [during the Q&A] and said: 'It is no thanks to the mayor or his splendid administration—NO! This is a farce. It was thanks to the battle that took place on the barricade last year on 26 March, it is thanks to that mayhem; it is thanks to those injuries as well as the injuries of the police that the government listened to our voices—it is not thanks to you, mayor. Who is he? There was already a mayor and an administration [when the companies arrived] and they are thieves.
>
> —Ray

In November 2014, the first wind energy Free, Prior and Informed Consent (FPIC) consultation began in Juchitán de Zaragoza, Mexico. Lasting eight months until 30 June 2015, this consultation responded not only to the United Nations International Labour Organization's (ILO) convention 169 that Mexico signed in 1990, but as Ray points out above, the widespread revolt against wind energy projects on the Laguna Superior in San Dionisio del Mar, Álvaro Obregón, San Mateo del Mar and Juchitán. Before the FPIC consultation began in the Isthmus of Tehuantepec (*Istmo*), 1,608 wind turbines were constructed triggering repeated barricades to stop wind projects and corresponding clashes with police, which accompanied a legal strategy and a court injunction in December 2012 to halt the Mareña Renovables project (Howe et al. 2015; Howe 2014; Smith 2012). The FPIC consultation under discussion was summoned for the Eólica del Sur (South Wind) wind project, named after the powerful south wind in the *Istmo* that attracted investment into the region. As discussed in Chapter 4, Eólica del Sur is the consortium previously called Mareña Renovables that sought to build 102 wind turbines on the Santa Teresa Sand Bar (Barra), and another thirty on the Pacific Ocean around Santa Maria del Mar. The barricade fighting, human rights groups, a legal strategy and 2012 court injunction led by Indigenous groups, however, was able to *officially* cancel the project, (that later rebranded itself as Eólica del Sur), and force the companies to relocate the project to another space between Juchitán and La Ventosa.

Based on observing consultations, analyzing transcripts and informal interviews, it is argued that even if unsuccessful, the FPIC was wielded by the government as a counter-insurrectionary device with the aim of pacifying opposition and legitimizing controversial development projects. FPIC consultations emerge as a 'soft' and enlightened approach to counterinsurgency that harnesses democratic notions of incorporation, self-identification and decision-making that conditions negotiations to create, what local human rights defender Lucila Bettina Cruz Velázquez has called, 'a bureaucratic trap', which represents and articulates a strategy of 'inclusionary control' (see Dunlap 2014; Dunlap and Fairhead 2014). Counterinsurgency is about maintaining legitimacy, mitigating and keeping conflict in its most manageable phase—'peace' (Dunlap 2014)—while strategies of inclusionary control are the art of integrating and de-escalating rebellious, insurrectionary and autonomous tensions arising from systemic grievances such as poverty, exploitation, state violence and the overall submission to the dictates of the global economy. The FPIC procedure is an attempt to channel rebellious tensions into 'constructive' negotiations and 'appropriate' channels mediated by a bureaucratic procedure that affirm state and corporate organizational processes and agendas, and by extension the trajectory of industrial progress and its social and ecological consequences. FPIC can be operationalized as a weapon that dispenses 'bureaucratic violence' in the service of fragmenting opposition and legitimizing controversial land deals (see Eldridge and Reinke 2018). With inclusionary control in mind, the following examines the FPIC consultation in Juchitán, Mexico.

This chapter begins with a literature review of the FPIC inquiry which also discusses experiences in other countries. Subsequently, I focus on the FPIC consultation in Juchitán, describing the consultation environment and efforts by government and wind company representatives sitting on the FPIC technical committee (TC) to frame and tailor their presentations on previous wind turbine grievances. The following section describes the tense and repressive situation during the FPIC consultation, and how people who expressed views critical and/or in opposition to the wind projects were targeted for intimidation, harassment and assault. Next, I analyze the contradictions within the FPIC inquiry as well as its inadequacies: not only in its procedures, but also in providing information about the financial, social and environmental impact of the wind project. This lack of information seemed intentionally misleading and rushed to fulfill ILO convention 169 requirements. The subsequent section considers the discursive techniques deployed by the FPIC consultation to manufacture the approval of the wind energy project on 30 June 2015. The final section concludes that the FPIC consultation undermines Indigenous autonomy, reinforcing a context of substantial political and economic asymmetry between state, corporate and elite interest and Indigenous fishermen and farmers that strengthens state power, and simultaneously serves as a marketing platform for development projects; thereby creating an illusion of real dialogue, negotiation and by extension democratic decision making.

Figure 5.1. 5 February 2015: Consultation at the House of Culture, Technical Committee during Q&A.

FREE, PRIOR AND INFORMED CONSENT (FPIC)

To fulfill the FPIC standards there must be *free*, non-coercive negotiations *prior* to any development intervention. This must also provide full and accurate information, and by extension establish an *informed* Indigenous population that have a complete understanding of the implications of the proposed project, allowing them to deny or *consent* to governments, national or transnational corporations to operate in their territory (Dunlap 2015; FPP 2007). Emerging from decades of struggle, Indigenous populations all over the world have been fighting for territorial recognition, rights and respect in the wake of colonialism, and in the face of state imperatives of modernization. In many ways, the FPIC mechanism epitomizes the legal outcome of decades of Indigenous struggle with colonial societies (Engle 2010; IACHR 2010), which is a mechanism that can be traced back to the decolonizing turn in international law in which Article 1(2) of the 1945 UN Charter established 'the principle of equal rights and self-determination of peoples' (Perera 2015, 147). This was followed by the 1966 Covenant on Economic, Social and Cultural Rights that protected the right to take part in cultural life, reaffirmed in 1975 by the International Court of Justice (ICJ). Where the ICJ specified the need for consent between states and Indigenous populations[1] and later gave rise to the 1989 legal document known as the Indigenous and Tribal peoples Convention, or Convention no. 169 of the United Nations International Labour Organization (ILO) (ILO 2009). Convention 169 came into force in 1991, affirming Indigenous rights and self-determination, embedding it in the FPIC protocol. Then the 2001 Universal Declaration on Cultural Diversity was adopted, which led in 2007 to the United Nations Declaration on the Rights of Indigenous Peoples (UNDRIP) and thus the reinforcement of Convention 169, cementing the importance of Indigenous self-determination to control their territory, lives and future (UNDRIP 2008).

The struggle for *Indigenous rights and self-determination* was embodied in both the UNDRIP and ILO 169 (ILO 2009), however, these conventions developed in concert with an ever present tension of what could be called '*financial realism*'. In 2002, the International Financial Corporation (IFC)—the financial arm of the World Bank—developed the Equator Principles: 'a risk management framework, adopted by financial institutions, for determining, assessing and managing environmental and social risk in [development] projects' (EP 2013). Commitments were made to corporate social responsibility and gaining a 'social license' to build a project, derived from the realization that failure to gain popular acceptance for a development project can lead to 'serious costs and delays' (Bakker 2013, 674). Said differently, gaining legitimacy, acceptance or now FPIC from local populations can mitigate project risks and unintended costs—delays arising from local opposition, security/mercenary costs or damages inflicted by recalcitrant locals during and after the project's construction. 'Social license' or corporate social responsibility (CSR) has the potential to promote, if not ensure, the smooth completion of projects and their projected revenue stream. This enlightened self-interest of the Equator Principles took another step forward in 2006 with IFC's Policy and Performance Standards on Social and Environmental Sustainability, which was revised again in 2011 to include FPIC (Baker 2013). Therefore, any development project financed or receiving a loan from the World Bank, Inter-American Development Bank (IDB) or the other seventy-seven signatory banks require the free, prior and informed consent of Indigenous populations (Baker 2013). IFC policy under guidance note seven, points to twelve states:

> *There is no universally accepted definition of FPIC.* [. . .] FPIC builds on and expands the process of informed consultation and participation described in Performance Standard 1 and will be established through *good faith negotiation* between the client and the Affected

Communities of Indigenous Peoples. The client will document: (i) the mutually accepted process between the client and Affected Communities of Indigenous Peoples, and (ii) evidence of agreement between the parties as the outcome of the negotiations. *FPIC does not necessarily require unanimity and may be achieved even when individuals or groups within the community explicitly disagree* (emphasis added; IFC 2012, 7).

FPIC is based on 'good faith negotiations', because it is not legally binding. Likewise, there is no one definition of FPIC, which does not require 'unanimity' or even a majority approval of people in affected communities to fulfill IFC criteria.

Inherently every aspect of FPIC comes into conflict with development projects (Baker 2013). While a *Free*—non-coercive—approach has to be judged within the context of the proposed project and *Prior* and *Informed* conflicts with up-front investments. Baker (2013, 693) explains, 'once the project developer has enough information to provide actual data regarding environmental and community risks to affected indigenous communities, the project will have already received its initial equity funding, a commitment that could prove difficult to unwind'. After a project is designed, approved for finance and environmental impact studies are produced, an Indigenous community's rejection of the project would result in the loss of hundreds of work hours and potentially many thousands of dollars. The reality of development finance, combined with not requiring unanimity or majority agreement with a project and, finally, the IFC definition of *consent as consultation* gives little substance to the protocol (IFC 2012; Baker 2013; Fontana and Grugel 2016). Defining consent as consultation consequently denies people the power to veto development projects—a power that is not mandated by Convention 169 even if it is often implied by the legislation. The IFC's interpretations of FPIC prove beneficial from the perspective of corporate social responsibility, 'framing FPIC as novel and gratuitous, both the banks and the international development community have created a mythology of justice' writes Shalanda Baker (2013, 686). From the perspective of Indigenous self-determination, this proves structurally deceptive and disingenuous. Nevertheless, the hope still remains for the FPIC protocol to strengthen Indigenous rights and exercise self-determination in the face of development projects.

The literature repeatedly stresses an unrealized potential in FPIC, while also elaborating on its difficulties with concerns about monitoring, certification, decision-making models, and the generalized disrespect of indigenous populations by governments and transnational corporations (Perera 2015; Doyle 2014; Franco 2014; Rotz 2014; Hanna and Frank 2013; Manhanty and McDermott 2013; Barelli 2012; FPP 2007). Similarly, FPIC is viewed as a mechanism to legitimize land acquisition, create social divisions and is a process of unequal power asymmetries (Flemmer and Schilling-Vacaflor 2016; Dunlap 2015; Fontana and Grugel 2016; Franco 2014; Rotz 2014; Baker 2013; Cariño and Colchester 2010). Jennifer Costanza (2015, 277) summarizes six complications to be expected with FPIC: (1) consultation will likely follow state or private company guidelines rather than Indigenous customs; (2) it might be unclear who is the affected community and what are their 'true' customs; (3) communities might not be fully informed about their rights and the proposed project; (4) consultation might begin after project approval; (5) an inadequate amount of time might be allocated to the inquiry; and (6) meetings might be scheduled at inaccessible locations. All these complications, except for inaccessible locations, emerge in varying intensities in the Juchitán consultation.

Despite a list of shortcomings, there are still possibilities for Indigenous populations to engage with FPIC. Jennifer Franco (2014, 5) summarizes the work from the International Institute for Environment and Development (IIED) that sees four positive possibilities. It can (1) help Indigenous people claim rights, resources and knowledge using national and international law; (2) strengthen the rules of communities to manage their natural resources; (3) create room for Indigenous groups to negotiate land deals, benefit sharing and access to resources; and (4) improve community cohesion and confidence to improve livelihoods and defend rights. In the case of Guatemala, Costanza (2015) pointed out the violent, racist and unfavourable situation faced by indigenous people where private companies hold eighty-four active mining licenses and the FPIC has failed to stop these concessions and only challenged one. Nevertheless, the consultation helped 'many indigenous people to rethink their identity, the rights they hold as indigenous peoples, and the role of the state in their communities; development and governance', where 'at least one' community altered governance practices and asserted land control over their territory (Costanza 2015, 277). Franco (2014, 17) also sees the potential for 'unintended and unanticipated impacts' that can be used to the advantage of people fighting for agrarian justice.

In Bolivia, Lorenza Fontana and Jean Grugel (2016, 250, 257) argue that not only is FPIC unlikely to 'resolve issues of democratic inclusion and participation,' but it opens 'up different kinds of political conflicts, between social groups and between society and state'. Further, FPIC is 'likely to become an exercise of strategic bargaining rather than an inclusive process with the "collective" interest at the center' (Fontana and Grugel 2016, 257; Manhanty and McDermott 2013). While Butzier and Stevenson (2014, 333) remind readers that FPIC is a 'legal obligation and good business', advocating a 'genuine and diligent efforts to seek affected indigenous peoples actual consent'. Jayantha Perera (2015, 160) concludes that 'FPIC, as a procedural law, brings some hope to indigenous peoples, particularly when state-sponsored resource exploitation and development policy ignore the legitimacy of indigenous institutions and interests'. The important conclusion is brought forward by Franco (2014, 3), who, while providing a comprehensive list of FPIC limitations, still asserts:

> FPIC is neither inherently 'good' nor inherently 'bad' from an agrarian justice point of view. Whether, how and to what extent FPIC processes can lead to outcomes that enhance agrarian justice will depend in part on the specific context in which they occur, and in part on whether and how pro-agrarian justice activist engage with them.

This argument is crucial. However, what if greater pressure is placed on FPIC? Franco (2014, 7) acknowledges that FPIC is increasingly used as a mechanism to facilitate and legitimate large-scale land deals and '(re)imagined as a tool for averting social conflict, while providing "social license" for deals to proceed'. This raises the question: When should FPIC consultations be abandoned and rejected if they are serving to continue an imposition against Indigenous autonomy and self-determination? Analyzing the FPIC in Juchitán, this chapter makes the argument that FPIC is a technique of 'inclusionary control' and 'bureaucratic violence' (Dunlap and Fairhead 2014; Kothari 2001; Foucault 1995/1977; Eldridge and Reinke 2018), intending to de-escalate resistance by incorporating the rebellious and insurrectionary tension inherent in the legitimate claims of Indigenous autonomy. Said differently, the mainstream popularity of FPIC[2] among international banks emerges from its counter-insurrectionary—counterinsurgency—potential to mitigate conflicts and costs to proposed development projects. In this view, FPIC is a political technology developed out of the real aspirations for self-determination which have been appropriated not only by the IFC, Inter-American Development Bank and transnational corporations, but also segments of the Indigenous population itself (largely their elites who will benefit and seek incorporation into the proposed projects) (Hall et al. 2015; Borras and Franco 2013; Borras et al. 2012). While FPIC is often the last hope of Indigenous populations to contest a project and assert their rights and territorial claims, it is also a key mechanism deployed when CSR fails to gain social license. After reviewing the FPIC in Juchitán, the conclusion will discuss the realities and difficulties faced by Indigenous communities and how to engage the FPIC consultation process.

LEARNING FROM THE PAST: FROM BARRICADES TO CONSULTATION

After ten years of wind energy development, the FPIC consultation began on 3 November 2014 and lasted until 30 June 2015. Located either at Juchitán Cultural[3] or Ecological Center, the consultation was arranged in a linear fashion, sitting in chairs or standing and facing the technical committee (TC), the members of which were to provide informative presentations and afterwards facilitate a Q&A session. The FPIC TC was predominantly male and composed of representatives from federal, state and international regulatory boards: The Mexican Secretary of Energy (SENER), Secretariat of Environment and Natural Resources (SEMARNAT), Federal Commission for the Protection against Sanitary Risk (COFEPRIS) and The National Commission for the Development of Indigenous Peoples (CDI). Professors from the National Autonomous University of Mexico (UNAM) and National Institute of Anthropology and History (INAH) also made guest appearances on the TC, alongside representatives from Eólica del Sur and the Juchitán administration.

Five factions could be identified at the consultation. First, members of resistance organizations such as the Popular Assembly of the Juchiteco People (APPJ) and the Assembly of Indigenous Peoples of the Isthmus of Tehuantepec in Defense of Land and Territory (APIIDTT). Members of these groups had been organizing against the imposition of wind energy since 2005, but the wind energy issue gained greater popularity around 2010 when wind companies began to prepare construction on the Barra and around the Laguna. Second, political parties, including the PRI, who held federal power for seventy years after the Mexican Revolution, and COCEI, a leftist party that formed in 1973 and combined Marxism with Indigenism. As noted in Chapter 1, after intense repression the COCEI began collaborating with the PRI by the late 1980s and subsequently began tolerating transnational corporations in the *Istmo*. Other political parties were also present, such as the Party of the Democratic Revolution (PRD), the Labour Party (PT) and the New Alliance Party (PNA) among others. All these parties retain links with key members of the COCEI.

The third group of actors identified at the consultation consisted of union leaders, landless and foreign workers associated with the PRI, COCEI and wind companies. The fourth involved transnational actors associated with the Spanish wind companies, representatives from international banks, particularly the IDB, as well as individuals representing and working with the Federal government. The fifth and final group comprised professors, students and NGOs including human rights groups that monitored the consultation such as CODIGODH; PODER; TEYPEC and Peace Brigades International (PBI). Notably, members of independent human rights groups, the APPJ and APIIDTT, were excluded from the TC.

The first consultation I attended took place on 4 February 2015. That day the ex-UN Special Rapporteur on the rights of indigenous peoples, James Anaya, sat on the TC to monitor the consultation. The Juchitán Cultural Center was filled with around 500 people and the situation quickly became tense. Numerous persons, looking like undercover police with ear pieces, held cameras and either pretended or actually had their hands on holstered handguns. There was a large crowd by the entrance that hunched over the chairs, creating an open space that also left chairs vacant. This crowd included *acarreados*—people paid to support political parties or unions. Once the Q&A section started, the distance between the crowd in the back and the chairs drastically changed. In one orderly and disciplined big step the crowd immediately moved forward, escalating tensions. The first people chosen for the Q&A immediately started insulting the wind company and the experts on the panel. By the end of the second Q&A speaker the crowd behind me started yelling: 'Liars, you work for the company!' A confusing statement given they were insulting government representatives. 'Out! Out! Out! Shut up, rat!' the crowd shouted.

Bettina Cruz Velázquez eventually got up and gave a counter-presentation explaining the problems with the consultation and the information provided about the wind energy project. This took twenty minutes. Meanwhile the situation became increasingly uncomfortable. I was in the middle of a group of people yelling: 'Out!' Drama ensued when the panel tried to shut down Cruz's presentation. The tensions rose significantly. I felt a brawl about to break out while Cruz was speaking. My friend began wrapping a paper clip around his finger to stab people when he punched them. He gave me one also, but I lost it somewhere; I felt it was inevitable that I or my friends were going to be attacked. This did not happen, and I was pleased a fight was avoided.

Bettina Cruz, a well-known human rights defender, is a member of APIIDTT and part of the National Network of Women Human Rights Defenders. Making the most of the Q&A session by giving a counter-presentation, Cruz discussed the private negotiations between wind company officials and government representatives in Huatulco, the failure of communal consultation, inadequate environmental impact assessments and the violation of Indigenous rights, while demanding that independent investigations be carried out on wind energy development in the region. Cruz's presentation triggered insults from parts of the crowd and the TC attempted to shut her down.[4] In general the consultation format seemed to promote a one-directional discourse, either from the panel to the audience or audience to the panel—there was no discussion. The speakers usually insulted members of the TC, politicians or other speakers on the panel in Zapotec, which was not translated by the Zapotec-Spanish translator. Similarly, questions asked by speakers about social and environmental impacts, income transparency, benefit-sharing and FPIC procedure were also noted. The TC sometimes commented on these concerns, but more often brushed them aside to be answered at the next consultation. Likewise, while there were concerns about unequal benefit-sharing and employment, union workers, land owners and wind company representatives advocated for the information phase to end and to proceed to the deliberation phase, during which construction could begin.

At the 3 December 2014 consultation it was clear that the government and wind company were trying to (re-)sell the wind project that was previously rejected on the Barra de Santa Teresa (Barra). With the past conflict in mind the Secretary of Energy Ramon Olivas called the transnational energy consortium Eólica del Sur a Mexican company, which is true in legal terms but deceptive in practice due to the company's extensive transnational composition. Eduardo Zenteno, director of Eólica del Sur, continued to emphasize the company's planned environmental mitigation as well as its reforestation programmes, endangered species protection, 'respect [for]... water resources', construction of roads in a way that will minimize flooding, giving community road use 'priority over any [construction] equipment' and the minimization of construction equipment noise. Furthermore, Zenteno announced the benefits of building new roads, giving preference to local workers for construction contracts and providing worker skill trainings. He also stressed that the wind company was not taking over the land but was merely leasing it, and promised social development and archaeological preservation programmes (Consultation, December 3). This list undoubtedly speaks to sources of conflict in the past that to this day remain points of contention surrounding wind energy development in the region. Nevertheless, a reputation of manipulation, broken and partially fulfilled promises left most people unfazed by these claims, while my friend went so far as to comment: 'He is just like every other fucker who sells you death'.

FREE? INTIMIDATION, VIOLENCE AND EMPLOYMENT OPPORTUNITY

The FPIC emerged after years of intimidation, targeted violence by the police, gunmen and even some instances of military intervention as well as full-scale barricade battles and shootouts in towns surrounding the Laguna Superior (see Chapters 2, 3 and 4; CODIGO 2014; Howe et al. 2015; Smith 2012). This continued during the FPIC consultation, with a tense and combative atmosphere with insults, public threats, intimidation and fights, sometimes during, but more frequently after the consultation. The people who I had been interviewing instructed me not to talk to them at the consultation, as it would also make me a target. Juan Antonio Lopez (CencosTV) summarizes the situation:

> There were aggressions in consultation sessions like attempts of physical aggression, insults, and coarseness, by people openly controlled by the [wind] company, they even delivered the sandwiches in the same place and outside they received payment for being there. These groups insulted the indigenous people who participated in the consultation. The sessions had harassment, gunshots outside some houses, blows to the doors, all these because of the participation in the consultation. Everything is in a report, the TC was informed and denunciations were presented for these aggressions.

Intimidation and opposition, according to informants, was funded by the wind companies directly and indirectly through political functionaries. During an interview in a neighbouring town a women explains how the *cacique* (political boss) pays people:

> So when people speak against wind energy [at the consultation], they will interfere and not let them speak. How do we know? We see the people there and the unions who want work.
>
> **AD:** *You saw the cacique give people two hundred pesos?*
> The guy who married her daughter [pointing to her friend] is the one who pays the people and sends them to Juchitán. The cacique hides, he is not on the front, and he sends people.

The cacique pays people 200 pesos to go and counter people in opposition to the wind park. The political elite, the *cacique*, acts as an intermediary between the wind companies and the town. The *cacique* are involved in politics, business and have clientelistic networks of supporters from distributing 'support'—money, food, work, buying votes and managing business in the town (see Chapter 2). A male participant explains:

> Even when we go to the consultation we see a lot of workers who are obligated to be there by the union leader. He brings them and gives them some money—the money comes from the [wind] companies, so they have to speak in favor of the companies because they are being paid to do that. Because businessmen do not want to lose the money and that is why they are in favour of these projects happening.

This individual continued to describe his first experience at the consultation. Seeking to speak out against the wind company because he lives near existing wind turbines, he described how he was confused by people from his town 'yelling and making noises' against people talking about the disadvantages of wind energy projects. The *cacique* managed the negotiations, contracts and social benefits from the wind companies which entailed unequal benefits distribution, and the available temporary work to be distributed to members of his networks. I was told that implicit in wind company work contracts is the informal obligation to support the wind projects, attend the consultation and invite friends, which can attract additional pay. This has become near public knowledge, not only from observing this at the consultation, where union leaders and wind company representatives manage the disruptions, but also with public radio confessions from workers and testimonies from family members above. Further, '[s]ome of the gunmen are working for the unions and they are defending their right to work', says a resident of a nearby town, while another woman contends that the *cacique* brings 'people with firearms to the front [of the consultation] all the time to intimidate people'. People interviewed repeatedly recount personal experiences with intimidation. Observations and testimonies attest that the political elites and wind companies in the region are bribing and buying people, sometimes armed, to disrupt speakers, attack opposition groups and manufacture widespread support in favor of the project.

Unions have blockaded highways around Juchitán, the town hall and even on the 14 April consultation they blocked the doors at the Cultural Center, holding people against their will for three hours. PODER (2015) documents thirty-two security incidents, which represent direct violations of the 'free' character of the FPIC consultation. The consultation was transformed into a low-intensity battleground to demonstrate approval or disapproval, which very quickly appeared theatrical given the procedural violations, outstanding information requests and an insistence to approve the wind project (PODER 2015; CencosTV 2015). As the excerpt from the field notes above demonstrates, the situation at the consultation was intense, conflictual and the FPIC consultation was surrounded by years of intimidation, shooting and battles with police to halt the construction of wind energy parks around the Laguna since 2012 (Introduction, Chapters 3 and 4).

PRIOR AND INFORMED

The 'prior' aspect of the consultation is about ten years too late. While many of the wind companies operating in the North of the *Istmo* claimed they have performed consultations, this has been refuted. There have been ten years of misinformation, disinformation, confusion and an overall lack of income transparency concerning social benefits and environmental impact, deceptive land lease agreements and intimidation for not signing or resisting these projects in the *Istmo* (see Chapters 2, 3 and 4). Likewise meetings regarding wind energy development were advertised poorly to the general public, focusing on land owners and, again, not the population at large, even though wind turbine construction would enclose entire towns and sit near homes, ranches and the sea where people fish (see Chapters 2 and 3). The prior aspect was at best selective and at worst non-existent.

Echoing the 'two broad types' of resistance—exploitation and dispossession—outlined by Borras *and colleagues* (2012, 413), the northern *Istmo*, now home to over 1,608 wind turbines, struggles with the effects of land change and steady increases in land, rent, gas, food and electricity prices, fighting for greater incorporation and collective social benefits for the town[5] (see Chapter 2). The southern part of the *Istmo* neighbouring the Laguna Superior and Inferior, after watching the situation unfold in the north, largely rejected the coming land change and terms of development while taking up a militant stance against wind companies to protect their land, sea and dignity (see Chapters 3 and 4). Meanwhile, the political elite and some land owners have benefited greatly, while other land owners benefit but remain dissatisfied with negotiations and worry about future environmental impacts and, if possible, continue to negotiate for more favourable contracts and hold out hopes for free electricity being possible for their towns.

This is the context that underlines the FPIC consultation, which makes the delivery of precise and detailed information of the utmost importance for the harmonizing of various local and transnational interests. The issues of unequal benefit-sharing, environmental impact, and social development are key and relate to race relationships—feelings of colonization by foreigners. Nevertheless, given the ten years of violations and internal conflicts over wind energy, it does not build support or legitimacy when there are known projections, negotiations and institutional approval for more wind projects in the *Istmo* (Rivas 2015). This point was stressed during the Q&A section of the consultation by advocates of Indigenous rights, summarizing this point Mariano Lopez on *CencosTV* (Aug. 20) states:

> [T]wo or four years before the consultation began, there were already rent contracts with Eólica del Sur which some indigenous farmers from Juchitán received to sign [contracts]. Besides there were two authorizations, one by SEMARNAT [Secretariat of Environment and Natural Resources] in EIA [Environmental Impact Assessment] in 2014, before to begin the consultation, and another by SENER [Secretary of Energy] in January 2015, during the Indigenous consultation process.

This violation of the Prior creates a weak, if not a disingenuous foundation of the consultation. In the same vein, the information provided ranged from unsatisfactory to insulting, depending on how emotionally vested a person was to understand the implications of the proposed project to make an informed decision. *First*, the linear layout of the panel informs people, while the people complain and ask questions that are systematically left unanswered, which spawns distrust, resentment and even hatred towards the TC. The TC gave people critical of, and who opposed, the wind park the feeling that the process was rigged, while people in favour of the wind park wanted to end the consultation and move to the next phase. While the consultations would last anywhere from two to seven hours (with the project already approved by Federal and state institutions), the consultations was reduced to pure theatrics to fulfill the legal requirement of Convention 169 to allow project construction.

Second, the collusion between the government and wind companies clearly shows that their interests are aligned and vested. Putting aside the years of combative struggle against the wind parks with state security forces aiding the operations to break Indigenous barricades against the wind projects, the government is responsible for economic, electric and climate change legislation promoting foreign direct investment (FDI), green economy development and by extension wind energy projects (see Introduction and Chapter 2). This has not only resulted in President Felipe Calderón inaugurating and cutting the ribbon at the Acciona Oaxaca II, III and IV wind parks, but also President Piña Nieto inaugurating the new Central Eólica Sureste I phase II on 2 March 2016 in the town of Asunción Ixtaltepec, thirty minutes' drive northeast of Juchitán. The technical committee consists of representatives from government institutions, local political parties and Universities who agree on the wind project, but not the specific terms and conditions. In Álvaro Obregón, Bettina Cruz Velázquez explains that the:

> consultation is a right we have and a right we demand, but it is wrong that the government is leading the consultation for their own benefit. This consultation is a bureaucratic trap, so they can manipulate and control the process in a way they see acceptable, because they are only consulting politicians and people who already agree with the [wind] project.

In the consultation, Cruz reacted to this by offering a counter-presentation during the Q&A to point out inconsistencies, contradictions and problems with the wind project. The vested interest in the FPIC consultation should be recognized as a structural procedural problem. The state is not a neutral actor or final arbiter—the state is a colonial force concerned with economic growth and its own organizational survival. This is a structural challenge that will always—to various degrees depending on its context—operationalize the FPIC consultation as a self-legitimizing mechanism for governments and their corporate partners to control, develop and convert Indigenous territory with or without direct opposition.

Third, inadequate information was provided. When the consultation completed the information phase on 30 June to begin construction—there were seventy-five unattended information requests and twelve neglected themes absent from the consultation, the latter of which the TC acknowledged (PODER 2015). The environmental impact assessment (EIA) was a key contention because: (1) SEMARNAT approved the high-impact project before the consultation; (2) the EIA was outsourced to a private company; (3) Indigenous people did not participate in the assessment; (4) the current EIA was regarded as inadequate for not or inadequately addressing issues around soil contamination, water table disruption, harm to animal populations and cultural sites, and so on; and (5) the study denied the existence of specific plants and trees in the project site. Demonstrating the depth of the concern

Mariano Lopez (CencosTV) explains: 'We asked for these studies because we have farmers with oil in their wells and this is due to his neighbour, who has a wind turbine thirty or forty metres away. We have asked for information about this type of oil and its toxicity, but the TC does not answer'. This was in addition to information about other health impacts such as distance, noise and electromagnetic currents—topics partially addressed in the 5 February consultation—which sold the public decibel levels as appropriate, minimized health impacts, talked about distances from houses that do not exist in most northern towns in the *Istmo*, and preformed an inadequate literature review. On 5 February, Silvia Victoria from COFEPRIS summarizes health risk evidence:

> *There is not enough evidence that says the infrasound emitted by wind turbine affects our vestibular system which is responsible for our balance.* There is not enough evidence suggesting a relation with vibrations and mental disorders or problems with mental health neither. The same is with the shadows we see when the wind turbines are moving, *until now there is not enough evidence to say there is a relation between the shadows and our cognitive and physical health.* Until now there is not anything scientific that helps us to say we are suffering this in our health because of the sound or infrasounds that wind turbines make (emphasis added).

Victoria claims 'there is not enough evidence' and the scientific literature is debated; therefore, it is appropriate to proceed with wind turbine construction. Yet there is a significant amount of evidence and research on health and environmental impacts from industrial wind turbines (Tabassum-Abbasi et al. 2014; Jeffery et al. 2013; 2014; Farbound 2013; Baker et al. 2012; Havas and Colling 2011). Ironically, Victoria mentioned home distances from wind turbines at 2.5km away and 'the recommendation is to watch them 1.4 or 1.5km away because it's dangerous to be under the wind turbines'. The entire towns of La Venta and La Ventosa are surrounded, while La Mata, Santo Domingo Ingenio and Juchitán are partially surrounded. These towns, rural homes and ranches are within ranges of thirty meters to 1200 metres. Comprehensive and independent EIA would likely cause outrage over existing wind turbines, denial of the Eólica del Sur project and/or new benefit sharing negotiations that would favour locals. People have observed the past six years of wind energy development in the north, and are aware to various degrees of the environmental impact on both humans and animals which has raised concerns and more specific questions.

Some landowners feel it is worth the cost, while others do not. On this, Baker et al., (2012, 48) writes: 'People who live close to wind turbines and do not benefit economically will be at risk to experience sleep disturbance and psychological distress. This risk increases with increasing sound levels'. Said differently, poverty, money and house proximity to wind turbines will help determine people's sensitivities to existing wind projects. However, the issues of wind energy should not be explained away by economic incentives, especially in situations of poverty. Otherwise low-income areas subject to wind energy or other projects become hostage to the imperatives of corporations and foreign direct investment (FDI), which stifles other more sustainable alternatives and developmental paths for local populations. A comprehensive, independent and participatory EIA could promote planning suitable to locals affected by wind energy or other megaprojects, which remained unrealized in Juchitán while the TC pressured the consultation into the deliberative phase with the help of select landowners, unions and the wind company.

Fourth are the technical violations and inadequacies present in the FPIC consultation procedure. This included lacklustre and limited notification and advertisement of the actual consultation. After complaints were made, there were improvements with wider media announcements on radio, in newsprint and on cars. However, this was combined with last-minute meeting changes with no or meagre notifications and to no effect. Similarly, the FPIC consultation would frequently start late, sometimes two or even three hours late. Making matters worse, occasionally the Zapotec/Spanish translator would not be present, alientating speakers of the dominant language of the region. This situation created additional consultation cancellations, stoking social tensions and imbuing feelings of disrespect with members of opposition groups. The cancelled 23 March 2015 consultation is a notable example.

The Prior and Informative aspects of this consultation were inadequate. Nevertheless, the discourse and techniques used were more than just neglecting information in wind turbine impact. This process was selling the wind project, creating a discourse of inevitability and, as can already be gathered, performing a theatrics to manufacture the legitimacy of continued wind turbine development. The next section will analyze the discourse of the FPIC and how it was constructed to legitimize the project even while failing to completely pacify opposition.

THE THEATRICS OF LEGITIMIZING DEVELOPMENT

Initially, the FPIC consultation was viewed as an opportunity for Indigenous populations to exercise their legal and constitutional rights. This sentiment was short-lived. At every consultation, it became increasingly clear that this was a legally-mandated procedure to approve and legitimize the continued expansion of industrial-scale wind energy projects in the *Istmo*. The presentation of the technical committee was one of marketing, mixed with an atmosphere of intimidation to opposition groups and people with genuine inquiries and questions. The divided interpretation of the FPIC consultation (noted above) between Indigenous self-determination and financial realism would manifest between local COCEI politicians, who discursively supported the possibility of rejecting the project, while government and wind company representatives painted a discourse of inevitability. The Secretary of Energy, Ramon Olivas, on 3 December 2014 explained to the consultation audience:

In general we are going to see the permissions for wind energy projects. SENER at the national level issues a self-supply permission; SEMARNAT at national level issues the project's MIA, the Ecology State and Sustainable Development Institute issues a EIA too for internal roads; the Directorate General of Civil Aeronautic (DGAC) for light beacons; SCT [Secretariat of Communications and Transportation] in Oaxaca delegation has to give permission for crossroads and access; and finally the local government the construction license.

The wind project had already received the government's stamp of approval; now the task was to perform a FPIC consultation.

For people living near wind energy parks, electricity was the first point of discontent. Ironically, because of the land change from agriculture to wind turbines, rising electricity rates had become a reoccurring problem with the arrival of wind energy in the *Istmo* (Chapter 2). Linked to the 1992 'self-supply' (*autoabastecimiento*) law that allowed corporations to buy their own wind parks and generate their own electricity, energy is owned by shareholders and exported to industrial zones, specific businesses and cities in Mexico, Guatemala and the United States (see Chapter 2; Oceransky 2011). Olivas attempts to normalize this controversial law by self-referencing government institutions, and how they already practiced this in Baja, California and Nuevo Leon—'this is not disadvantageous for the country according to SENER'. The message is clear: the government approves it, therefore it is normal, people should not contest wind energy or how energy is used. The crowd is then reminded: 'Oaxaca is the third state at national level with the highest subsidy for energy', implying that people in this area should feel privileged for the current state of energy affairs. Olivas's talk was dedicated to minimizing the very real discontent of rising energy costs, rural gentrification and poverty entrenchment that is compounding work-related out-migration. Not to forget, this energy is being exported to other countries—US, Belize and Guatemala—and to corporations such as Walmart, Grupo Bimbo and industrial construction companies (see Chapter 2). Then, people are told in the consultation that this situation will improve with the next wind park.

This narrative of normalizing is reoccurring. During the 5 February 2015 consultation dedicated to health impacts, the first twenty minutes of the presentation was dedicated to explaining the production of wind energy around the world. Not only does this appear as filler, but the message is clear: China, the USA, Germany, Spain and India are generating electricity, so Mexico and the *Istmo* have nothing to worry about and should be doing it as well. This deflects time away from people's more pressing worries to discuss the complicated debates surrounding wind energy health-impacts, and it is framing the discussion in a way to normalize and promote a project rather than to discuss the real trials, tribulations and unknowns regarding health impacts. This neglect of health issues is matched by parading and promoting the discourse of the green economy—'with this project it is possible to avoid 879,000 tons of greenhouse gas'. How does this resonate with people who live surrounded by wind turbines, facing increasing electricity, land, rent and food prices as well as facing health problems that many are convinced are produced by their proximity to these enormous wind turbines? The further people are from living under wind turbines, the more effective the discourse of the green economy becomes—even in towns surrounded by them. If people do not feel the direct impacts of the development, then they tend to look away and show disinterest, especially if they have a monetary vested interest in the projects.

While the panel was largely marketing wind turbines, many questions were raised. Systematically, the points about income transparency, social, health and environmental impact were ignored, stalled or avoided. Repeatedly, after February, the panel would tell people regarding environmental impact statements: 'We will deliver this document today; at the end of this assembly you can go to the table and take it'. The people who conducted the study were unable to be present at the consultation because they were out of the country. Again, it was an old EIA report that did not answer many of the pressing questions regarding noise, headaches, concrete foundations, leaking oils and so on—all extremely sore points for locals living, farming and fishing near wind turbines.

The TC would criticize people for not asking direct questions. However, people would not only use the Q&A as a time to make a speech and insult the TC or for those in favor of the project opposition groups, but people also systematically asked for information concerning health and environmental impacts. This was in addition to multiple invitations offered to the TC by residents, to come and see wind turbine distances to houses, oil in water wells and the condition of animals. The questions raised concerning wind turbine income transparency law, health, social and environmental impacts were never answered by the TC. Instead the consultation was filled with summaries rushing the consultation into the next phase and repeating the same information regarded as unsatisfactory by opposition groups, which fuelled an exhausting Q&A that would go on for two to four hours refuting the information presented, insulting the panel and FPIC procedure. Similarly, when people questioned the consultation for procedural violations, inadequate information and cultural violations the TC responded: '[W]e have to remind you that this is an unprecedented process here in Juchitán, that started with good faith from authorities to respect what Convention 169 says, which is the instrument that regulates the consultation' (March 25). Novelty and gratitude became an excuse for inadequacy, disrespect and ignoring threatening disruptions.

The FPIC consultation did not fulfill its anticipated potential to affirm Indigenous rights and self-determination. The promotion of the wind project, incomplete health information, and failure to conduct adequate social and environmental impact reports, as well as concerns around income transparency, discredited the FPIC consultation in Juchitán. The TC was committed to rush to the deliberative phase regardless of thirty-two security incidents, seventy-five unanswered information requests, and the TC's own acknowledgement of twelve unanswered information themes that all combined to create a theatrics to legitimize wind energy development. '[T]his hurry the authorities have', said Mariano Lopez (CencosTV 2015), 'is because of some deadlines that financiers are giving to the Eólica del Sur in order to begin the project'. The project was already approved by government. Private negotiations were taking place in parallel to the FPIC consultation and the TC, with a conflict of interest, was pushing through the FPIC procedure in clear violation of Convention 169 and in the face of popular protest.

Regardless of the discursive manipulations, project marketing and cultural insults, on 30 June 2015 the information phase would come to a close with outstanding procedural violations, open information requests, and the ever-present intimidation. This was formally protested by resistance groups, human rights organizations, journalists, NGOs and factions of the COCEI, which accompanied a legal strategy. After collecting 1,166 signatures, on 15 September 2015 opposition groups filed a court injunction (*amparo*) with the Seventh District court of Salina Cruz (Manzo 2015). On 30 September a temporary injunction suspended all permits, license approvals and land change approvals to Eólica del Sur that, later, on 16 December 2015 became a permanent suspension (Sin-Embargo 2015). Then on 9 June 2016, the Seventh District Court denied the injunction (Orozco 2016), creating an opening for continued wind park construction in Juchitán County. The legal success in exercising Indigenous rights against the Mexican government, Eólica del Sur and the FPIC procedure was short-lived, which is a struggle still in process.

CONCLUSION

Therefore merely by attending the meeting they had muzzled themselves, bound themselves to new bosses who were more insidious than the old bosses because they came from among themselves.

—Sophia Nachalo, *Letters of Insurgents*

An exercise in democracy, incorporating and creating a new arena to exercise Indigenous rights, the FPIC consultation in Juchitán was employed to approve and legitimize the Eólica de Sur wind project. The FPIC consultation in Juchitán failed to provide genuinely *free*, non-coercive information *prior* to approving the project and thus did not provide the conditions to allow for *informed consent*. Instead the FPIC consultation in Juchitán all but mimicked the negative features mentioned in the literature review and suffered a conflict of interest between state and corporate collaboration, violated cultural norms, provided inadequate information and served as a marketing platform to sell the idea of the project rather than adequately addressing the issues raised by participants, such as income transparency or social and environmental impacts. Consequently, the FPIC has served to entrench state power and reinforce existing political and economic asymmetries between different actors. Interestingly, in this case, the Mexican courts proved more accountable in responding to Indigenous rights and raising questions regarding the FPIC's operational purpose and its ability to affirm Indigenous self-determination.

Battles against state security forces and struggles for Indigenous autonomy directly and unapologetically challenged the legitimacy of the state, its formation and continued development. Governments and corporations classify these struggles as 'insurgent', regardless of their legitimate justifications. Behind every FPIC consultation is the practice or threat of military, police and other security forces occupying Indigenous territory to enforce colonial law and protect corporate investments. FPIC consultations represent a 'soft' and enlightened approach to combating insurgency that harnesses the democratic techniques of incorporation, self-identification and participatory decision-making for the purpose of 'inclusionary control'. The FPIC represents a relatively new mechanism that repackages old labels, procedures and paperwork to integrate people into a bureaucratic (colonial) system that provides hope, enables (real or imagined) expression/feedback and regiments people to chairs, presentations and Q&A proceedings. Strategies of inclusionary control work to facilitate a shift within rebellious groups from *total rejection to negotiation* which inherently relies on the threat of coercive force and works to create spaces to facilitate or widen existing social divisions. This is the strategic articulation of bureaucratic violence, a colonial violence used to fragment resistance to controversial development projects and legitimize the conquest of territories by industrial infrastructure and everything they symbolize. Counterinsurgency fashions the FPIC procedure as a concession made attractive by the everyday political and structural violence that wears people down and creates openings to lure recalcitrant populations into rigged democratic theatrics that tightens the psychosocial grip of state and corporate power.

Despite discursive and physical attacks, the resistance remained determined in quickly recognizing the intentions of the FPIC consultation. From the perspective of Indigenous autonomy, the mere attendance of the consultation promotes submission to procedural disciplines, state authority and acceptance of land acquisition via subterfuge. Thus, FPIC is not a neutral political technology of deliberation and decision-making but a device of social mediation and control. It emerges as a liberal colonial technology with vested interests and power asymmetries and functions as a component of the state apparatus bound to the imperatives of political control and economic growth, which incidentally are said to be the foundations of ecological crisis and anthropogenic climate change (Dunlap and Fairhead 2014). In short, it might be beneficial to view the FPIC consultation as a politico-military hold-and-build technique designed to (re)establish control and legitimacy over populations resorting to direct action and asserting their legal rights against destructive development projects. If the possibility exists, rejecting FPIC consultations might be wise until the process is taken seriously by granting impacted people veto power, eliminating corporate-state-elite interests from them and enforcing the actual meaning of the words *Free, Prior and Informed Consent*. Oppositional groups need to consider the best ways in which to use their time and resources to accomplish their goals of defending their ecosystems and ways of life, which might include boycotting FPIC consultations and expending their energies elsewhere.

NOTES

1. Further, there is a colonial history of extermination, assimilation and homogenizing difference between Indigenous people into a manageable political category, which also denies the recognition of many Indigenous nations, creating a politics of state recognition (Hale 2002; Churchill 2003; Moses 2008; Woolford et al. 2014).

2. This is the cultural equivalent of Payment for Ecosystem Services (Dunlap and Fairhead 2014).

3. This is a location that the COCEI struggled for during the 1970s and thus retains important symbolic meaning.

4. Cruz's presentation time count: 30:30–53:00, available at https://www.youtube.com/watch?v=0qfbJ9QDrrM.

5. Political corruption is a problem multiplier in this regard.

6

Renewing Destruction: Colonization, the Genocide-Ecocide Nexus and Wind Energy Development

> The road to hell is paved with good intentions.
> —English proverb

> There are ways to destroy people's lives without killing them.
> —Hex, Jormungand

When I first arrived in the Isthmus of Tehuantepec region of Oaxaca, Mexico to research wind energy development, I knew there were a range of issues emerging over their construction (see the Prologue). I would soon find myself, however, confronted with a discourse that was far more intense than I originally imagined. This emerged in conversations and interviews with people claiming that the Mexican state and wind companies 'are going to kill us all', 'annihilate us' and assertions that they are slowly committing 'ethnocide', 'ecocide' or 'genocide.' After two weeks of living in Álvaro Obregón, this perspective first emerged in an interview with an anarchist who had been visiting the town for over two year. Discussing the different perspectives on wind energy in the town, they described an argument at a party with a Constitutionalist (Contra) who believed wind turbines were an opportunity for the town and the gateway to 'progress'. Then I asked:

> **AD:** *You are someone who isn't from an 'Indigenous community', per se; you grew up with more of the luxuries of modernity than some of the people in this village. So when you hear him [the Contra] say: 'You are a halt to my progress, you have no right to do this', which implies that you have modern luxuries that you are preventing him from having and that you have no right to prevent him from having those luxuries. What do you say to this?*

[. . . .] When they tell me I am an impediment to progress and that I have no right to do this: First, I believe that this issue in not just [local]—I hate the term local and I think this term comes from neoliberalism, nobody is a local. It has a reference that reflects the metropolis. But what is happening here is happening in a lot of places in this country and in this state and it needs to be halted. Because I think the progress of modernity is a threat to life itself—it is going to kill us. It is a hegemonic power that does not respect any other way of life and it has to be stopped. The Industrial revolution in Europe destroyed a shit-ton of things there, it wiped out just about everything—a bunch of species, trees and many other ways of living. So yeah, it has to be stopped and it does well fucking concern me! But I think the power to halt the renewable wind energy projects is important, because yes, I believe these projects are part of a strategy of ethnocide—I do think that they want to kill them [the Indigenous people].

I did not expect this reference to ethnocide. Also, in that moment, it did not dawn on me to integrate questions about this perspective into the semi-structured interview questions, instead continuing to focus on asking about social impact, conflict dynamics and repressive strategies deployed by the Mexican government and wind companies. References to extermination, however, recurred throughout interviews in the region, and given the circumstances I experienced in the area, after fieldwork I began to inquire into colonial genocide studies to see what insights might support this argument. This chapter emerges as a product of this research, examining the relationship between wind turbines and genocide.

The chapter argues that wind energy development advances a trajectory of progress that requires political and ontological assimilation that amounts to continuing (see Blaser 2013), what Jennifer Huseman and Damien Short called, 'a slow industrial genocide' (Huseman and Short 2012). Wind energy is renewing the destructive tendencies of industrial development in the coastal Isthmus of Tehuantepec region (*Istmo*). Driven by increased international/national emphasis on renewable energy and market-based environmentalism, wind energy necessitates not only the intensification of enclosure and privatization, but also renews direct and indirect coercive impositions on Indigenous territory and land relations to further integrate wind resources into capitalist production. Beginning by discussing theory, this chapter examines the colonial model, the political state and its relationship to the 'genocide-ecocide nexus' to contextualize how wind energy development continues the trajectory of colonial genocide by advancing social and environmental pressure on Indigenous groups in the *Istmo*. The first section begins by outlining a definition of colonialism that assists in identifying the temporal continuity of the colonial project to later understand its relationship with wind energy development. The next section provides a literature review on colonial genocide studies, which briefly discusses the increasing relevance of self-management in colonial genocide studies, the 'genocide-ecocide nexus' and the 'intent' of destructive development projects. This leads into reviewing the claims and findings that emerged from fieldwork in the *Istmo*, which is divided into the North and South to show the different, yet similar dynamics taking place in the region. Finally, the chapter concludes that wind energy development as a 'solution' to climate change (as it is positioned next to more overtly destructive methods of energy production— oil, hydraulic fracturing, coal and nuclear energy) not only distracts from its dependence on fossil fuels and mining, but renews and advances the colonial genocide process by continuing to assimilate and target (Indigenous) people who continue to value their land, culture and relationships.

WELCOME TO HELL: THE COLONY MODEL

Colonization is not a discussion of the past, but the present. This requires an examination of the colony model and its evolution to better understand its continuation and relationship to climate change, ecological crises and wind energy development. Dating back to the Roman Republic, colonialism is often wedded to the Latin word *imperium*—imperialism—that signified Roman states, while the colony originates from the word *colonia* that designated Roman military settlements on conquered territories (Moses 2010/2008). The Roman Empire had a robust agricultural system to support its territorial expansion (Roth 1999), which was incorporated into military camps, stimulating agriculture production in their regions of settlement (to feed soldiers) to the extent that there is evidence of soldiers farming (Southern 2006). Integral to the Roman camp were roads to enable military mobility, making the legionaries' skilled both as soldier as well as road builders (Thompson 1997). The relationship between the military camp, roads and agriculture is deeply intertwined and is the heart of the colony model in Roman times and afterwards.

Imperium and *colonia* would combine to create the notion of Empire. This signified the domination of one society by another by military force, designating the different tactics, strategies and politics of creating and maintaining an empire (Moses 2010). Hence the imperial relationship of unequal exchange that refers to the country-colony, centre-periphery and the urban-rural divide that often implies dependency, if not a type of master-slave relationship. The 'recurrent problem' concerning the relationship between 'colonialism' and 'imperialism', explains Daniel Butt (2013, 892), is that some scholars see colonialism as an instance of imperialism, the domination of a territory from an external metropolis or nation, while others refer 'to a particular model of political organization', which emerges from the Roman military camp. Quoting Edward Said, A. Dirk Moses writes, "'imperialism was the theory, colonialism the practice of *changing the uselessly unoccupied territories of the world into useful* new versions of the European metropolitan society", others simply equated the two' (emphasis added, Moses 2010, 22). This resonates with the claims of developers calling land 'available', 'unproductive' and 'underutilized', implying the need to transform it into an infrastructural or industrial agricultural project (White et al. 2012; Schutter 2011, 543), this definition locates the core of colonialism, which can be synthesized as the assertion: that there is only one right way to use land, live, organize culture and/or develop a nation and that is through techno-capitalist development. Inherent is a sense of superiority that articulates itself not only through overt domination with the 'right of conquest', but also the good-intentions that manifest in paternalism, charity or as one critics have called 'the white-man's savior complex' (Cole 2012) which can even take the form of 'solidarity' from activists (see Anonymous 2014).

Colonialism, according to Moses, is 'a specific form of rule' that is embodied by the 'occupation of societies on terms that rob them of their "historical line of development" and transforms them "according to the needs and interests of the colonial rulers"' (Moses 2010, 22). Consequently this results in the great transformation of Indigenous social, cultural and economic institutions, which disciplines ontological outlooks, interpersonal relationships between humans and, equally important, non-human life or 'more-than-human nature' (plants, animals, landscapes, trees, etc. (Abram 1996; Sullivan 2017). Nonetheless, Butt (2013, 892) identifies three primary characteristics of colonialism: (1) the external domination of one people by another; (2) the imposition of colonial 'culture and customs onto the colonized'; and (3) the exploitation of the colonized (slavery, natural resource extraction and 'misappropriation of cultural property' to name only a few. This is also justified what Gayatri Spivak (1988), following Foucault (1989/1961), called epistemic violence, which constructed a method of knowledge to justify claims to superiority over the 'Other' (see also Marker 2003). Such a method simultaneously served to discredit and subjugate alternative perspectives, ontologies and knowledges that asserted different values and ontologies (Foucault 2003). While these are foundational characteristics of colonialism, if one wanted to understand the heart, composition and the way colonialism enforces these cultural relationships, then these definitions remain relatively open-ended and ambiguous.

In order to understand and identify the colonial model and its relationship to wind energy, Paul Virilio (2008/1983, 166–167) is helpful when he writes: 'the colony has always been the model of the political State, which began in the city, spread to the nation, across the communes, and reached the stage of the French and English colonial empires.' Said another way, colonization is the processes which spread a form of organization emblematic of the ideology, form and purpose of the European political state. The spread of which Lorenzo Veracini (2014), following Patrick Wolfe's distinction between colonialism and settler colonialism, argues that colonialism takes on viral and bacterial qualities (see also Rahnema 1997). Revealing the affinity between the Roman camp and the political state, Foucault (2007/1978, 15–16) explains:

> The form of the Roman camp was revived at the end of the sixteenth and the beginning of the seventeenth century, precisely in Protestant countries—and hence the importance of all this in Northern Europe—along with the exercise, the subdivision of troops, and collective and individual controls in the major undertaking of disciplinarization of the army.

The Roman camp, laid the foundation for a 'military dream of society' (Foucault 1995/1977, 169), designed around divisions of labour, specialization and hierarchy, and later adopted by European states to reproduce these values to regiment people to imperatives of nation-state formation, 'modernization' and later industrial progress. Anarchists call this 'prison society' (RF 2013, 7; Weir 2016). While rooted in the Roman camp, the colony begins in the city, but more specifically, the ancient Greek *Polis* and organizational precursors in the ancient Greek *oikonomia* (household governance) (see Arendt 1998/1958; Owens 2015). Demonstrating the affinity between the colonial/state model and democracy, Ferit Güven (2015, 99) returns to Plato's question in the *Republic*: 'What is the best way to rule a community as a whole?' Democracy would emerge as the answer, distributing the crisis of central authority and the desire for political unity in the Ancient Greek *Polis* (small community organized around a town centre), establishing citizens (-slaves), civil participation and the nascent juridical-political machine of the *Polis* (Güven 2015, 6). While there are many forms of democratic participation, democracy as a system of political control[1] arises from how to construct and/or maintain the *Polis* (and *oikonomia*), exemplified by Attica as an organizational structure of participation that united spatial and political practices that served as an inspiration for the Roman camp. Investigating democracy as a political disciplinary technology, Güven (Güven 2015, 4) defines colonialism as 'neither a simple series of acts of domination, nor an unqualified exploitation, but rather a process and discourse of disciplining, ordering, rendering visible, unveiling, and making comprehensible'.

Güven's definition does not explicitly discuss spatial qualities. This 'disciplining, ordering, rending visible, unveiling, and making comprehensible', however, is indeed coded into colonial space and organizational practices. Achilles Mbembe (2003, 26), summarizing Franz Fanon, explains that:

> colonial occupation entails first and foremost a division of space into compartments. It involves the setting of boundaries and internal frontiers epitomized by barracks and police stations; it is regulated by the language of pure force, immediate presence, and frequent and direct action; and it is premised on the principle of reciprocal exclusivity.

Re-designing space to promote spatial legibility for population control, economic imperatives and industrial growth is foundational to nation-state development and re-development that codes its values into space (Foucault 2007; Rodgers and O'Neil 2012; Dalakoglou and Harvey 2012). Enforcing the colonial model with military, even genocidal force, occupation and administration established an evolving and dynamic system of conquest. Restricting colonialism to a historical era appears short-sighted, naïve and a politically convenient conception to enable the unchecked expansion of the colonial model, an expansion we can call political economy. This structural, organizational and technological relationship not only situates the evolving nation-state, but contextualizes the historical place and operation of wind energy.

COLONIAL GENOCIDE: REVISITING THE GENOCIDE MACHINE

By defining colonialism, it enables us to locate the continuities and changes taking place within the colonial system and its relationship with genocide, development and as we will see later with industrial-scale wind energy parks. The politics of genocide studies has been contentious, grappling with the legal politics of the 1948 United Nations Convention on the Prevention and Punishment of the Crime of Genocide (The UN Convention), Raphael Lemkin's original definition of genocide as well as issues of 'intent to destroy' as they relate to economic operations. Following the assertion that the process of wind energy development in the *Istmo* is continuing that of colonial genocide, this section engages in a literature review to understand genocidal processes to contextualize and substantiate this claim.

The word genocide, coined by Raphael Lemkin in the 1944 book, *Axis Rule in Occupied Europe*, combines the word *genos* meaning tribe or race of people and the Latin *cide* meaning killing (Hinton 2002). Genocide is tribe killing, which Lemkin originally described as 'a coordinated plan of different actions aiming at the destruction of essential foundations of the life of national groups, with the aim of annihilating the groups themselves' (Lemkin 2005/1994, 79). This includes 'the disintegration of the political and social institutions of culture, language, national feelings, religion and the economic existence of national groups, and the destruction of the personal security, liberty, health, dignity, and even the lives of the individuals belonging to such groups' (Lemkin 2005, 79). Furthermore, Lemkin (2005, 79) writes:

> Genocide has two phases: one, destruction of the national pattern of the oppressed group: the other, the imposition of the national pattern of the oppressor. This imposition, in turn, may be made upon the oppressed population which is allowed to remain, upon the territory alone, after removal of the population and the colonization of the area by the oppressor's own nationals.

Destruction and reconfiguration of cultural values, or deterritorialization and reterritorialization (Deleuze and Guattari 2005/1987) are notable characteristics of genocide which only partially appear in the 1948 UN Convention. Article II states 'genocide means any of the following acts committed with intent to destroy, in whole or in part, a national, ethnical, racial or religious groups as such:'

(a) Killing members of the group
(b) Causing serious bodily or mental harm to members of the group
(c) Deliberately inflicting on the group conditions of life calculated to bring about its physical destruction in whole or in part
(d) Imposing measures intended to prevent births within the group
(e) Forcibly transferring children of the group to another group (Hinton 2002, 43–4).

The negotiations over the terms of the UN Convention were largely conditioned by US and Soviet concerns of self-incrimination for Indigenous extermination and assimilation (Moses 2010). Furthermore, the UN Convention did not recognize political identities;[2] produced a dominant conception of genocide based on mass-killing; and, most importantly, largely neglected aspects of 'cultural genocide' in Lemkin's writings (Short 2010).

The politics behind the UN Convention, Lemkin's wider definition of genocide and the continuing destructive processes of capitalism, the nation-state and industrial development have given rise to an ontological split in genocide studies. Moses (2002) makes the distinction between liberal and post-liberal conceptions of genocide. The liberals (Drost 1959; Glaser and Possony 1979; Bauer 1982; Rodger 1987; Chalk 1994; Markus 2001; Quigley 2009; Behrens and Henham 2013; Kirsch 2013) emphasize state actors, intentionality, totalitarian ideology and the industrial-killing emblematic of the Nazi regime, which they position as historically 'unique', 'unprecedented' and 'singular', consequently downplaying and, at times, positioning colonial genocides as insignificant compared to the atrocities of the Nazis (Short 2010). The post-liberals on the other hand (Césaire 1950; Arendt 1951; Sartre 1964; Davis and Zannis 1973; Clastres 1994/1980; Foucault 1998; 2003; Perlman 1985; Barta 1987; Churchill 1997; 2003; Palmer 2000; Curthoys and John Docker 2001; Evans and Thrope 2001; Hinton 2002; 2004; Abed 2006; Wolfe 1999; 2006; Powell 2007; Shaw 2007; Padel and Das 2010; Short 2010; 2016; Woolford et al. 2014; Leven and Conversi 2014; Kingston 2015), see genocide as more complex than *just* mass-killing, highlighting not only the differences, but also the continuity between colonial genocide(s), the Holocaust and how these acts are coded into the structure and trajectory of the nation state and its development.

The focus for post-liberals is on how states continue to eliminate and/or reconfigure competing value systems into its institutional and economic structures. 'When I say "Killing", I obviously do not mean simply murder as such', said Foucault (2003, 256), 'but also every form of indirect murder: the fact of exposing someone to death, increasing the risk of death for some people, or, quite simply, political death, expulsion, rejection, and so on.' Foucault's notion of killing here is in line with the post-liberal conception(s) of genocide, where 'indirect murder' references the under-acknowledged concept of 'social death' that describes the hollowing out of cultures, habits and religions. Social death does not directly kill people, but instead disciplines and transforms them, instilling various degrees of helplessness, social fragmentation, extreme depressions and post-dramatic stresses, among other existential crises (see Short 2016). Damian Short (2010, 842) reminds us that 'social death can occur without specific 'intent to destroy' as such, through sporadic and uncoordinated action or as a by-product of an incompatible expansionist economic system. They might even result from attempts to do good: to enlighten, to modernise, to evangelise.' Social death calls attention to the science of elimination, assimilation/conversion and population management that has been normalized within the state apparatus, its social (schools, malls, hospitals, public space, etc.), coercive (military, police and extra-judicial adherents), and cultural (churches, museums, family structure, etc.) institutions.

The liberal position has been rightfully critiqued for a type of Eurocentric exceptionalism placing greater importance on the Holocaust (Moses 2002), 'invoking a snapshot view of history divorced from past context and experience' that prioritizes legal definition over lived experience (Short 2010, 840). The post-liberal approach rejects the narrow definition of the UN Convention, drawing connections between colonialism, economic development and genocide, which consequently discredit the foundational myths of liberalism (Short 2010). Furthermore, the post-liberal position argues that this process is intensifying with technological advancements to prefect systems of production, consumption and resource control, constructing what amounts to a 'genocide machine' (Davis and Zannis 1973). Keeping the idea of social death in mind, discussing democracy Güven (2015, 11) writes: 'While the totalitarian regimes "take power by destroying all oppositions", within democracy opposition are allowed to survive' as long as they conform to the rules, operations and imperatives of political economy. On the other hand, post-liberals are criticized for not providing culpable agents and being static in the face of changing dynamics (Moses 2002), which, this chapter contends, misses the nuance, flexibility and adaptation of the colonial system in its ongoing normalization of violence, maintaining its existence and integrating opposition into its structures—leading some authors to call capitalism structural genocide (Jones 2000; Leech 2012; Kingston 2015; Wolfe 2006).

This liberal critique neglects the important conversation concerning the internalization, self-identification and reproduction of colonial values. It should be clear that not only is a colonial/state pattern imposed (referred here as political economy), but it can be self-managed by the targeted population with a series of disciplinary and biopolitical mechanisms that constantly works to integrate people into its political structures (Dunlap 2014a; Galeano 1997; Güven 2015). Wolfe (2006) sees colonial genocidal processes in three non-deterministic phases: (1) initial confrontation (or invasion); (2) carceration period (displacement/resettlement); and (3) assimilation period that aims to integrate indigenous populations into the colonial system.[3] Here, a fourth phase should be added: *self-management*, which is an intensification of the assimilation phase to normalized colonial structures, making them self-reinforcing and managed. In short, state/colonial structures seek to become socially and economically *sustainable*. The colonizing process, also known as 'modernization', 'industrialization', and 'development', intensifies and continues to reconfigure the most sensitive features of people's cultural values and sociality, amounting to a systemic social war to regiment and integrate people into the various operations and (managerial) positions within the system of political economy (Dunlap 2014a; 2014b; 2017; Galeano 1997; Güven 2015).

Neoliberalism names the present and ever evolving structure of conquest, where Marx's (2010/1887) primitive accumulation and Harvey's (2005) accumulation by dispossession become useful theoretical tools that chart the process of cultural genocide that integrates people directly (through overt force) and indirectly (through assimilation practices) into the economic system of the state, which takes on worldly proportions and links, under the idea of globalization—a world system (see Escobar 2004). Self-management, then, not only attempts to erase past genocidal and internment programmes, but it also works towards normalizing present forms of symbolic, structural, infrastructural and political violence (Bourgois 2001; Rodgers and O'Neil 2012). This legacy of violence severely complicates political analyses, which frequently remains neglected by mainstream social science. This analytical neglect arguably results from various and overlapping degrees of self-identification, dependency, addiction and/or desire for the social, ecological and self-destruction implicit with the colonial/state systems. William Dugger (1988, 92) reminds us:

> Social control through coercion is temporary. More permanent social control is based on the ability to alter the internal values of others to gain their willing acceptance of the control. Then the control becomes legitimate. It is deemed right and good by those over whom it is exercised. It no longer requires a whip.

Almost everyone, to varying intensities, is disciplined, integrated and inculcated into the machinations of colonial structures that construct an organizational system that promotes self-domestication into institutional, environmental and cultural structures that blur individual social values, self-interest and identities with the functioning of political economy itself. 'In other words, to become a colonizing culture, Europe first had to colonize itself', explained Ward Churchill (2003, 236) to a German audience, '[i]n fact, your colonization has by now been consolidated to such an extent that [. . .] you no longer even see yourselves as having been colonized.' This alludes to the erasure and/or normalization of genocidal violence, the importance of colonial collaborators, the manufacturing of consent and/or desire to emulate colonial culture. In sum, genocide impacts and influences everyone, underlining nearly every political and ontological ecological conflict wedded to the implementation of destructive development projects.

Before moving forward, there are two important insights from colonial genocide studies that should be mentioned. First is recognizing the relationship and inseparability of Indigenous people and their land, where ecologically destructive interventions are then experienced as attacks against Indigenous (and non-Indigenous) populations subsisting, valuing and identifying with their ecosystems (Blaser 2013). Developing this insight, Crook and Short (2014) recognized ecocide as a method of genocide that forms the 'genocide-ecocide nexus', which attempts to undermine the life, existence and resistance of Indigenous populations[4]—where killing the buffalo, fish, crops and other means of subsistence are textbook counterinsurgency 'starvation' tactics part of a larger extermination strategy (Isenberg 2001; Churchill 2001; Boot 2013). Said simply, attacks against the land can have genocidal consequences for (already marginalized) Indigenous communities, groups and individuals who derive their material and spiritual life from the land, which was mentioned briefly in the conclusion of Chapter 3. Second, regards the 'intent to destroy' and development projects. Severe physical, cultural and ecological destruction resulting from mega-development projects are disregarded and denied, because the 'intent' underlining this destruction is 'economic'.

The resulting death, displacement, illness and social fragmentation are then seen separately, distanced from pervious and existing processes of physical and/or cultural genocide and, instead, understood as 'unintended consequences' of the project (Maybury-Lewis 2002, 70). Short (2010, 835), following Helen Fein, contends that genocidal 'intent can also be *inferred from action*, which is a long established principle in British common law.' Judging by the action and outcome as opposed to legal script, assessing the genocidal-ecocidal impacts as an outcome rather than based on the standards, visions and laws of state institutions—a product of colonial conquest themselves—is a step forward in breaking the colonial spell, maybe even 'decolonizing' approaches to genocide studies. The genocide-ecocide nexus is a long-term, continuous and coercive process operating by various means and methods. The key point here is that, in the words of Wolfe (2006, 388), '*invasion is a structure not an event,*' imbuing discipline, desire and coercion to make the colonial/state process self-managing and reinforcing.

CULTURAL GENOCIDE AND WIND TURBINES

The wind parks and their continued expansion south had become an increasing source of discontent in the *Istmo*, especially for those who continue to appreciate their subsistence from the land and sea. The specificity of how the wind turbines entered into the different parts of the *Istmo* had both difference and commonality, but in the present struggle the North and South coastal *Istmo* represent two different archetypal, yet overlapping, forms of resistance that are revealing of their respective contexts. Opposition in the north is centred on unequal, exploitative land deals and labour contracts as locals fight for greater incorporation, as well as for individual and collective benefits. This includes unions fighting for more wind parks, yet which also criticize the wind companies for the importation of technical employees and unequal pay between Mexican and Spanish workers. While in the south there is the total rejection of wind energy projects, largely arising out of the belief that the wind companies and political system cannot be trusted and propagate lies to take their land and damage the sea—their ability to subsist.

Resistance in the north was scant and fragmented because of a pre-existing concentration of political power, land ownership and selective dissemination of information to the general public—many people had no idea what wind energy was, the projects' scale or their social and ecological impacts (see Chapters 2, 3, and 4). Additionally, individuals had ambitions concerning profit, while the town was wrapped in hopes of social development and projects were promoted as environmentally sustainable or 'green' and thus work towards mitigating climate change. Eventually,

La Venta and La Ventosa became engulfed by wind turbines, while other towns such as Santa Domingo Ingenio and Juchitán were only partially enclosed.

Interviews in the north dripped with discontent. This ranged from manipulation of land contracts to unrealized social development and to what amounted to a type of rural gentrification that made nearly everything in towns go up in price. This also included the exporting of electricity from these wind parks to industrial zones, mining companies and other countries while electricity prices increased locally (see Chapter 2). People were engulfed by electrical infrastructure and in some instances people live between distances of thirty to 280 metres from wind turbines both in and outside the La Ventosa and La Venta. Further, during interviews people discussed the land use change around the town that altered agricultural and livestock patterns and necessitated the clearing of animal habitat, compacting of soil for roads, loss of bird, turning the ground water into concrete for wind turbine foundations and finally, leaking oil into the ground, which people claimed contaminate both the ground water and animals. Albeit less extreme, these ecological impacts are similar to other modes of conventional fossil fuel energy production (Downey et al. 2010; Huseman and Short 2012; Short 2016).

According to local famers, wind turbines and their foundations create extreme drying and flooding of the land depending on the season, which made it difficult to continue farming. This was compounded by various reports of minor-to-severe health impacts that resonated with the controversial wind turbine syndrome. As Chapter 2 discussed, these findings resonate with other studies on wind turbines impact. Despite some wealth increases and token social development projects in the last ten years of wind energy development in La Ventosa, these projects appear to have largely reinforced income inequality, furthered poverty entrenchment, increased food vulnerability and worker dependency on the construction of more wind parks, cumulatively leading to an increase in work-related out-migration and environmental degradation (see Chapter 2). The presence of wind turbines creates a struggle to make the best out of a bad situation that leaves people fighting for more social benefits, hoping for landowners to negotiate free electricity for more wind projects, as well as adapting to the situation to survive. One woman summarized the situation saying: 'We are still poor and now we are surrounded by wind turbines.'

The South coastal *Istmo* on the other hand, witnessed and listened to the stories of wind energy development in the north. Politicians and some landowners were interested in negotiating the terms of the Barra de Santa Teresa (Barra) project and communal land outside Juchitán. The process began with the help of local elites, politicians and interested *comuneros*: the general public was initially left in the dark, later taking a position of total opposition. Public consultation was bypassed, instead opting, in both the north and the south, for selective negotiations with select regional administrators, elites and social property members. This resulted in the spread of social conflict all along the southern towns, notably San Dionisio del Mar, Álvaro Obregón, San Mateo del Mar and Juchitán, between the years 2011–2015, a struggle that continues to varying intensities today. It took over ten years for the first Free,

Prior and Informed Consent (FPIC) consultation to arrive in Juchitán, which, as Chapter 5 discussed, reinforced state-corporate power, while simultaneously acting as a wind energy marketing platform and constructing the illusion of real dialogue, negotiation and, by extension, democratic decision-making.

The key difference between negotiation/incorporation with wind companies in the north and the current of insurrection against them in the south is that these were fishing communities dependent on the sea, not only for material, but also spiritual sustenance—identifying as part of the sea. This raised deep concerns about the environmental impact of wind turbines and how the construction, vibration and noise from the wind turbines would affect aquatic life from which they live. Likewise, land concentration was more diffuse: aside from *ejidos* there is also the existence of communal lands and roads, which wind turbines threatened. Similarly, political corruption, unequal land deals and a loss of access to the sea was combined with fishermen witnessing the mass-killing of fish during the construction of a pilot wind turbine on the Barra. This led villagers to unite and rise up against the wind companies and political parties as outlined in detail in Chapter 4.

The quality of life, land relations and livelihoods were threatened by the possibility of construction on the Barra and the Lagoon. The Barra is comprised primarily of sand and vegetation and the Mareña Renovables project sought to build 102 wind turbines with the foundation depth estimated up to seventy metres deep as opposed to the average eight to thirteen metres deep, on the land in the north. The first attempt at building a foundation, according to testimonies, resulted in the mass-killing of fish as far as the eye could see (see Chapter 4). In addition to wind turbine construction, barge docks, a less than one kilometre submarine transmission line and a fifty-two kilometre transmission line to Ixtepec substation would be constructed (IDB 2011). This would create a situation of systematic noise, vibration, electrical currents and aircraft warning lights flashing on the wind towers, which the fishermen in the Seventh Section neighbourhood in Juchitán will tell you, pushes the fish populations deeper into the lagoon. This means fishermen have to drive to other towns to fish, fomenting inter-communal conflicts when other towns or members thereof collaborate with the wind companies, but then travel to other villages resisting wind energy development to fish. Drawing from these accounts and secondary literature on wind energy impacts, these wind projects represent a structure of systemic low-intensity[5] ecological destruction which the majority of the residents in the south felt had to be stopped at all costs. Hence the emergence of militant resistance against these projects in Juchitán, Álvaro Obregón and the lesser mentioned San Dionisio del Mar and San Maria del Mar.

Cultural Change: The Northern Coastal Istmo

After having been enclosed by wind turbines, people in the north live next to or are at the centre of wind energy generations sites. This has resulted in cultural change, which is compounded by a rise of health affectations—real and imagined—and widespread reports of cancer, the source of which remains undetermined. The changes to these towns have been significant. The findings suggest that a dramatic rise in land, rent, food, electricity prices that paralleled an increase in drug consumption and crime accompanied the wind energy projects. This included a large and rapid influx of wealthy foreigners, migrant workers and their preferences. Likewise, wind energy development is proclaimed as green, sustainable and climate friendly, but in reality wind energy development is still the result, in every aspect of its production, of fossil fuels. Wind energy still requires large mining and processing facilities to refine iron, stainless steel, dysprosium, oil lubricants, sealing resins, fiberglass and concrete as well as the construction of transportation and electrical infrastructure networks (see Dunlap 2018). These issues are compounded by the fact that the electricity generated by the wind turbines is private under 'self-supply' (*autoabastecimiento*) regime that reserves the energy produced for shareholders, such as Grupo Bimbo, Walmart, industrial construction companies and mining companies, and is exported to the US, Belize and Guatemala. The energy from these wind parks do not go the residents and is largely controlled by transnational companies.

The difference between residents who participated with the wind companies and those who did not or could not, resulted in a rise of income inequality which was further exaggerated by increases in land, rent and food prices. This manifested itself on the street in La Ventosa with infrastructural degradation, signs of malnourishment in people and raggedy clothes that coincided with the circulation of brand name American SUVs, new clothes and refurbished or new-build compound-style homes with fresh coats of paint, tiles, barbwire and sometimes security cameras. These infrastructural trends combined with changes in food. Referring to a restaurant in La Ventosa, one local human right activist explains that:

> you could eat garnachas there, cocada, torta, coffee—not anymore, it is all gringo food. Light-skinned people like you, more or less, go there and they have their menu. This is a way of understanding how the people from [the wind companies] there are modifying their way of life and that is without thinking about the economy that revolves around them.

Large influxes of foreigners with money brought their habits, lifestyles and preferences with them, which subtly altered the type of food and price of food in the town (Chapter 2). Similarly, one woman believed the severe health problems in the town were linked to greater dependency on the importation of food; people used to eat free range chickens, pigs and cows, but had now become more reliant on canned food from neighbouring regions. People in La Ventosa also repeatedly spoke about a rise in crime and drugs, which some are convinced emerged from a lack of opportunity, wealth inequality and the foreigners associated with the wind companies bringing new lifestyle habits. Consequently, the locals claim this has generated greater insecurity in the town, while ironically, stories circulate about the police protecting the wind park and actively arresting and fining people enormous sums of money for hunting. Hunting has been an important part of people's livelihoods and seasonal festivals, both of which are being abolished by police enforcement of these laws, while protecting wind turbines that are repeatedly cited as destroying animal habitat, killing birds and, according to research participants, reducing the overall animal population. The prohibition of hunting is another cost placed on people who are disadvantaged by the arrival of wind energy in La Ventosa.

Interestingly, there is a form of settler colonialism taking place, organized around megaprojects. Influxes of European businessmen, representatives, engineers and other workers from around the world flood these small towns for about one and a half to three years. Not only does this influx bring promises of prosperity, jobs, social development and images of modernization that create enchanted visions of development (see Harvey and Knox 2012), it also seems to be changing the prices in the town, the amount of drug use and social composition at an accelerating rate. The town is flooded with people holding values which prioritize wind park construction, capital generation and the right to work. This takes on intimate qualities when workers and wind company representatives begin dating and marrying into the town not just in the north, but all over the coastal *Istmo*. Numerous research participants felt young women were taken advantage of through promises, drugs, money as well as seduced by older, confident and 'blue eyed' foreigners.[6] According to interviews, once the wind projects were complete there became a lot of 'fatherless children' as people moved on to the next job and/or went home to their families in other countries. While victimization was a common narrative, it was also said, that many Istmeños women were active, if not strategic, about forming relationships with foreigners to further their love, life and possible opportunities. Power is fluid, flexible and worked to the strengths of each individual; nonetheless, these relationships also created openings to communal integration and access to land in the area. Said simply, marrying into the region provided access to land and wind resources as discussed in Chapter 2; a relationship that is by no means simple, but allowed for the further penetration of wind energy and the change that comes with it.

The circuit of labourers that travel from megaproject to megaproject—renewable or otherwise—forms a roaming settler-colonial machine that integrates the values of modernization based on the demands of work. In La Ventosa, after the wind company jobs left, there was a rise of mototaxi drivers, landless workers and subsequent out-migration creating, after Marx, an increase in the 'industrial reserve army' for the semi-specialized labours on the megaproject construction circuit, plus workers in seasonal agricultural, factory, tourism and construction, among a variety of other available jobs (see Green 2011; Stephen 2007). Megaprojects in general, and wind energy in particular, creates a self-fulfilling cycle of state, economic and energy dependency on infrastructure systems through large-scale development that gradually weakens localized food systems. They also promote development on a scale incompatible with ecological sustainability, which stifles potential developmental alternatives in the process (Sale 1991/1985; D'Alisa et al. 2014; Escobar 2015; 2018). The end result is a continuation of cultural change along the lines of the neoliberal vision that is increasing political conflict, altering productive structures, and promoting gated community-style homes, SUVs, commodity pricing, food systems, drug consumption and health concerns (see Laurell 2014).

The Southern Coastal Istmo

The south is subject to similar changes from environmental impact to the influx of wealthy foreigners, but these issues are significantly compounded by wind energy development on the sea. Not only is the sea a significant cultural symbol,[7] but also foundational to livelihoods—'We are poor, but you do not die of hunger here', as one person explained. Arguably, fishing creates a greater connection to the cycles of the ecosystem, but also provides an immediate correlation between life and the sea. This is not just with fishing, but also with direct subsistent food systems, where dependency is closely situated with the land and sea as opposed to bureaucratized market-based supermarket food systems.[8] Despite centuries of colonization, bombardments from media advertisements, movies and 'the merchants of cool' (Goodman and Dretzin 2001; Bernays 1947; Brock and Dunlap 2018), as well as new schooling regulations that many feel are eroding traditional values in the areas around Álvaro Obregón (Chapter 4), a large portion of the population remain steadfast to defend the land, sea and their cultural integrity from wind companies and their political collaborators. Two foundational issues emerged with wind energy development in the Southern coastal *Istmo*: changes in food production and quality of life.

The land change from agriculture to wind turbines—which according to farmers also accompanies cuts in agricultural subsidies—threatens their existence. In the words of one farmer outside Juchitán: 'If [. . .] the wind turbines arrive, there are not going to be farmers anymore and that is when the natural food will be finished'. Preserving living and working with the land is crucial to the inhabitants I interviewed and the arrival of wind energy threatens to disrupt and, as is claimed by many, significantly degrade their quality of life. Summarizing this relationship, a farmer explains: the wind company 'is going to give you money so you can eat food from over there [the city]—canned food. Who knows what canned food is!? Sometimes the food is over eight months or a year old, not like the food here in Juchitán'. This farmer, and others interviewed, became enthralled talking about the diversity and quality of food in the region, and consequently resents the imposition of urban food systems with the quality of life it brings. The same farmer continues:

> I do not throw fertilizers onto my produce, only pure nature, because the food has a lot of medicine. [For example,] [t]he soda that has a lot of gas, makes the people die faster. All of my family, my grandpa and my grandma they lived to 105 and 115 years old. They were drinking [hot] chocolate at six in the morning, they ate chocolate every day, cheese and tortilla, but real tortilla's from the oven. Now tortilla stores use machines and machines use gas to cook, which is why people's hearts are stopping and they die, but the real tortillas that came from the oven they used wood, that is why the people lived to be 110, 115 years old. Now the people who do not eat fresh food, they die at fifty or sixty years.

This antagonism towards processed food, from industrial fertilizers to canned food, was related to life expectancy. This is a relationship that is being threatened with the arrival of wind turbines, which according to this farmer is made worse by money. 'Many of my friends who received money, they say they live happily now, but they drink [beer] all day and every day and one of them has already died because of this money'. Money is implicitly described as a weapon before the farmer drifts into a tangent about the great quality of tortillas, tamales and *atole*,[9] later explaining that the people who sign contracts with the wind company drink all day and are 'dying because they are assholes'. Someone from the Communitarian police in Álvaro Obregón explains that the wind turbines will destroy everything, 'even their watermelon, which is what happened to people in La Ventosa. We want to continue free, so everybody can work on their farm, have their crops, farm their corn, beans, squash—all of that—and we can continue to fish'.

Remembering 'Big Bear' in Chapter 4: the life brought by wind turbines, many felt, was going to drag them into a form of wage slavery.

From the inheritors of revolutionary agrarian reform, modern life looks like slavery: people work all day, have little money and no time to enjoy life. This type of life is common in urban areas, but regardless of material poverty in the coastal *Istmo*, people still share their food and eat high-quality fish, fruits and vegetables. Another farmer explains: 'There is only going to be people with money, so they are going to eat canned food, but our sons are going to suffer a lot, like the elementary school books say: "That many people work from six to six and they have chains on their feet", this is coming back.' These older farmers and fishermen associate modernity and wind turbines with slavery, which they claim comes at the expense of a few getting rich while the majority get poorer, meanwhile farming and fishing conditions are degraded and the existing problem of work-related out-migration only increases. Furthermore, acceptance of wind energy development in this region is viewed as putting their kids' and grandkids' futures at risk—limiting their freedoms, quality of food and relationship with the land and sea. On the contrary, people in favour of wind turbines feel they are putting their children's futures at risk by not bringing money and social development, or having them fall behind the Mexican standard of education. There is an active and fervent desire to acquire greater incomes, move away from subsistent living and ride that one-way train of Rostow's (1960) *Stages of Economic Growth* with wind energy development being the ticket.

These are some of the feelings leading to popular revolt against wind energy development. The people revolting against these projects are no doubt many and have legitimate concerns about these projects, but as is common with discrediting resistance, from colonial times until now, the opposition groups are slandered as a 'minority' of 'violent' and 'drunk' bandits. This is not to say all these claims are entirely untrue, but this is the propagation of a moral discourse to delegitimize the concerns of people seeking to protect the land and sea and to preserve what remains of their cultural integrity. The impact of wind energy is substantial, as are the qualitative dimensions and ontological relationships with the earth, food production and how life is lived by the primarily Zapotec residents.

Wind Energy and the Genocide-Ecocide Nexus

Taking the land in the name of sustainable development is accumulation by dispossession by environmental ethic—green grabbing—which walks a fine line with genocide. 'The Rana' exclaims: 'We hold responsible all of the political parties of Mexico, the government in its different levels for the attempt to annihilate us, the attempt to grab our land and to wipe us off the map'. The elder Mapache, discussing the struggle of the community council of Álvaro Obregón against the Mexican government and wind companies asserts that as 'long as we can hold out and have people's support, we will say: "You will have to kill us first."' With tears running down his face, he continues by saying that the community council will defend their territory and 'that is why they are going to kill us'. The struggle over wind turbines is conceived as a war devised to 'annihilate' them, which is conceived as a generational fight—'for my people, for our sea, our land, our children, grandchildren and future generations'. Assessing the benefit of the wind energy projects, the 'Wild Tiger' asserts that 'the wind energy project for us is a tool of ethnic cleansing that does not build anything useful for me'.

Delving into the genocide-ecocide nexus and relating it to offsetting (see Sullivan 2010; 2017, Brock 2018), the land defender, 'Hada' explains:

> [T]here are over two hundred types of medicinal plants and each medicinal plant has its area—its natural habitat. So when the wind energy projects invade they kill this area, they kill part of the herbs. How are they going to transplant those herbs to another place that is not their natural habitat? [. . .] The same with the animals: they already have their dens, they already have their special trees where they make their nests—where they reproduce. So if you watch the bird eggs, not all of the birds are returning to hatch them, it is a really high risk of extermination. I call it ecocide, genocide that is what they are doing with our way of life. Water, which is our vital liquid, [. . .] to build the wind turbines they are digging six to ten metres under the ground, which they fill with cement and rebar, and so it brutally harms our life.

For the Zapotec and Ikoot people listening, interacting and identifying with the land, they see their fate intimately related. Describing the impact of wind energy development on natural medicines, Hada demonstrates a facet of the genocide-ecocide nexus that is not only targeting plants, animals, water and people, but inhibiting people from accessing and engaging alternative health practices. Wind companies deny that they are causing this type of harm (GNF 2013b) and claim that farming can co-exist with wind turbines,[10] while corporate social responsibility (CSR) offers social development schemes and offsetting programs utilized to justify this industrial destruction (see Brock and Dunlap 2018). Wind turbine development is destroying natural medicines, killing fish, impeding cultural sites (see Chapter 3) and altering land relationships deeply tied to the livelihood and culture of Zapotec and Ikoots people. This, however, continues by technical, subtle and indirect means. When farmers refuse to sell their land, as mentioned in Chapter 3, their fields are then enclosed by wind turbine access roads 'raised about sixty centimetres' which consequently transforms their farmland into a pool during the wet season. This is a type of intentional, or worse, unintentional environmentality (Foucault 2008; Galbrys 2014), that organizes space to be flooded, causing Indigenous people who value their land and agrarian lifestyles to be displaced into the homes of relatives, urban job markets, and work-related out-migration corridors, or otherwise be both literally and figuratively drowned. This method of flooding deployed against 'stubborn' or determined farmers imposes new values and land relationships prioritizing wind energy development while disregarding the agrarian relationships and culture, meanwhile displacing Zapotec, Ikoot and others further into a position of dependence, if not submission, to a neoliberal political economy.

This situation is intensified by overt and covert methods of repressive counter-insurgency techniques. The good intentions of the green economy are combined with the violent intent to secure investment. The factions totally rejecting wind energy developments perceive this struggle as not only about wind turbines, but also about how to preserve and continue of what remains of Zapotec and Ikoot life ways 'that value mother earth', 'know how to ask for forgiveness' from the land and to share harvest yields.[11] While there are benefits to out-migration (Stephen 2007), local experiences with work-related out-migration suggest severe hardship during travel, language difficulties, racism, abusive working conditions, and shifts towards excessive drug use, leaving people feeling trapped as disciplined workers and consumers between the extremes of village and urban life. Negative experiences with migration leaves the impression that people now value their cultures, connection to the land and semi-subsistent lifestyles even more, which underpins resistance to wind energy projects.

Wind energy development is an opportunity for modernization, economic growth and carbon dioxide reduction. Carbon dioxide reduction assumes the industrial economic trajectory, which does not challenge but only 'softens' industrial degradation by deploying the logic of econometrics to justify what amounts to a double-bind of the political lesser of two evils between fossil fuels and renewable energy. This leaves the root cause of these problems unchallenged. Despite this logic, that does not challenge the industrial-scale degradation of human and non-human natures, the intention of the economy is 'good' by way of seeking to 'develop' the 'underdeveloped', making useful 'unutilized' land and, in the words of the Inter-American Development Bank (IDB 2011, 10) on the Mareña Renovables wind project: this is 'land exposed to intense human activities in the past decades which have led to a deterioration of the "natural" character of the area'. This means, given that the 'natural character of the area' has deteriorated and given the situation of material poverty, wind energy megaprojects, like all foreign direct investment (FDI) according to the logic of neoliberal capitalism, is 'good' for (economic) development and will not make the environmental and social conditions any worse. This is *not* the case. The widening of income inequality, a large influx of foreigners, rising social conflict, disruptions to farming land and marine life as well as increases in land, rent, food and electricity prices are new or significant intensifications of existing social and ecological degradations. According to neoliberal ideology, FDI is always 'good', and as the CDM document states, concerning the Mareña project, it will: (1) develop renewable resources; (2) enforce environmental sustainability, avoiding fossil fuels; (3) generate employment for construction and maintenance; (4) maintain land owner income 'without giving up stockbreeding, fishing and agriculture'; (5) raise foreign capital; (6) diversify the national energy portfolio and (7) support infrastructure improvement (Roads, bridges, etc.).[12] These are the selling points, which from the perspective of political economy are assumed all 'good' improvements, progressive steps and measures to address climate change, and in the words of President Peña Nieto, help one of '*the most backward regions of the country*' (Paley 2015, 6).

The change taking place, however, is summarized well by Hada in Chapter 3, explaining the qualitative environmental changes on their land where the Bíi Hioxo wind park is now built. 'Before you went to a ranch [. . .] you hear the birds sing, the growling of the wind, all of this relaxes you, but now it is not that way. You go there to the rancho and you start hearing a bothersome Buuzzzzzzzzzzzzzzzzz—how can you relax?' From birds chirping and tranquillity to the buzz of electrical currents and wind turbine gears grinding, this is the change taking place. For people living with and connected to the land, the impact of industrial wind turbines is significant to say the least; for urban dwellers this might already be their life, but the monotonous buzz, turbine rotations and shadow flicker continue as long as the wind blows. George Tinker's (1993, 5) words come to mind when reading the CDM document above: 'the good intent of some may be so mired in unrecognized systemic structures that they even remain unaware of the destruction that results from these good intentions'. This quote is painfully applicable to market structures, whose ideological dogma is normalized to the point of justifying the continuation of a slow industrial genocide which further imposes capitalist values, land relationships and infrastructure into the cultural life of Zapotec, Ikoot and other farmers and fishermen in the *Istmo*.

CONCLUSION

This chapter has sought to investigate and develop the argument that wind energy is continuing a slow industrial genocide. The first section begins by outlining a definition of colonialism that assists in identifying the temporal continuity of the colonial project to understand its relationship with wind energy development. I subsequently asserted that the increasing relevance of self-management is key to the genocidal process, while also highlighting the 'genocide-ecocide nexus' and the 'intent' of megaproject development—green or otherwise—is to transform the land, displace or consume the human and nonhuman resources inhabiting the area. Following this, the key dynamics and outcomes of the 'wind rush' in the Isthmus of Tehuantepec region were reviewed. Listening to Zapotec and Ikoot experiences asserting the genocidal and ecocidal consequences of wind parks, it was argued that wind energy development continues a slow industrial genocide through market-based environmentalism and climate change mitigation programmes.

Interestingly, however, are the social divisions over wind energy. Not all Indigenous people are against wind energy in the *Istmo*. Resistance arises specifically from farmers, fisherman and others who recognize the intrusive colonial behavior, unequal benefit-sharing, disregard for public consultation, and the cultural and ecological impacts of wind parks. There are many perspectives regarding, and alternatives to, development, which include aspirations for communal and micro-scale wind energy developments directly linked to towns as opposed to profit-centred exporting of electricity to industrial centres. Despite the risk of essentialism associated with the label of 'Indigenous' people, there is an undeniable process of manipulation and coercion inherent in wind energy development in the *Istmo*. This process is legitimized through capitalist mentalities and growth imperatives via federal, state and local politicians as well as elites. It should be no surprise that capitalist culture—or Scott's (1998: 4) 'high-modernist ideology'—continues to dominate, which builds from and intensifies from earlier colonial, state and economic interventions in Mexico, and the *Istmo* in particular. Wind energy takes on genocidal qualities when flora, fauna and cultural relationships are being destroyed and/or re-regimented into 'offset' sites or migration corridors leading people into monocultural fields, tourist and industrial zones in Mexico and the United States. At issue here is the elimination of different cultural values, ontologies and relationships emblematic of Indigenous people's communion with the land and sea. The values and relationships of actively respecting and living with the land and resisting statist and market assimilation are barriers to 'development' and the specific targets of the slow industrial genocide advanced by wind energy development. The colonial process and its socially and ecologically destructive trajectory are self-managed by various people and identities—Indigenous politicians, elites and various peoples—which has long been the case in Europe and across the world.

The green economy thus emerges in the shadow of conventional fossil fuel production, presenting itself as a 'solution' and pathway to slow the effects of ecological, climate and economic crisis. Said differently, renewable energy in particular, and the green economy in general, emerge as the 'lesser evil' of industrial development. Discussing the principle of 'lesser evil,' Eyal Weizman (2011, 10) writes: 'less brutal measures are also those that may be more easily naturalized, accepted and tolerated—and hence more frequently used, with the result that a greater evil may be reached cumulatively'. The green economy is the lesser industrial evil, utilizing a technique of war to morally buffer and continue the proliferation of industrial waste in the name of climate change mitigation, which according to this research results in greater cumulative social and environmental alterations and even the systemic and increasing destruction of alternative value systems and ways of life valuing relationships with ecosystems. This is a process that is not separate, but builds, from processes of colonization, nation-state formation as well as energetic systems ranging from coal to nuclear power. Wind energy remains the least destructive fossil fuel energy technology, but this does not change its subtle and embedded logic of extermination that renews and extends the industrial system, consequently applying further pressures on the plants, animals and Indigenous (and other) people living from the land, sea and wind in the *Istmo*.

NOTES

1. See conversations: CrimethInc. "From Democracy to Freedom." CrimethInc, http://crimethinc.com/texts/r/democracy/; Invisible Committee. *To Our Friends*. Los Angeles: Semiotext(e), 2015; endnote 51, Dunlap, 2014b; endnote 16, Foucault, 2003; Ellul, Jacques. *The Technological Society*. New York: Vintage Books, 1964 [1954].

2. This transformed total and internal wars into 'Dirty War', as opposed to politically motivated genocide.

3. This process is also outlined in Foucault's (2003, 145) analysis of ancient conquest (see endnote 51, Dunlap, 2014a, p. 59/6).

4. M. Crook and D. Short, "Marx, Lemkin and the genocide-ecocide nexus," *The International Journal of Human Rights* 18, (2014): 298–319. See also endnote 40.

5. This is compared to processes of coal, nuclear, oil and hydraulic fracturing.

6. One among many testimonies explained: 'When the wind turbines began to operate, foreign workers began to arrive. They were specialized people, engineers and other specialists and they took advantage of the opportunity to sexually abuse young women. Because they were tall with blue or green eyes and they had money, they took advantage of the opportunity to sexually abuse girls of fourteen, fifteen and sixteen years old.' Interview, 18 May 2015.

7. This extends to ceremonial and relational practices, as well as the technomorphic references to the sea as a bank account, to try to communicate with financially minded foreigners.

8. Farmers have come to enjoy past agricultural subsidies, but these have slowly been eroded and this has helped facilitate the arrival of wind energy projects in the region.

9. A local hot drink made of corn, with many varieties.

10. GNF-(Gas Natural Fenosa). "Mitos Y Realidades" Online, 2013b. *Parque Eólico Bíi Hioxo*. Ed. Fenosa, Gas Natural. Blog at wordpress.com. 28 Jan. 2015. <https://biihioxo.wordpress.com/ecologia-y-cultura/>.

11. For example, Hada explains: 'So I learned from my father to value mother earth, how to ask for forgiveness and share the harvest. He farmed for birds, ants, widows and orphans. His first five furrows were to share and then the rest are for the family. And before planting my father would ask for forgiveness from Mother Nature, the land, and ask God if he could harvest what he planted. Digging a furrow was considered a violation of mother earth—that is why he asked for forgiveness and he would ask for a blessing for the harvest'. Interview, 13 March 2015.

12. CDM. Clean Development Mechanism: Project Design Document—San Dionisio Wind Farm, 2–3. Online, 2012b. UNFCCC: https://cdm.unfccc.int/filestorage/1/B/Z/1BZCF3LKGIUN4R2SATXV6Q7E8MJ9W0/San%20Dionisio%20Wind%20Farm%20-%20PDD%20resub.pdf?t=dWd8bzI0ZTN4fDB1unLi4PAZxayPaaebA-iq.

CONCLUSION

The Grid System Spreads, Dependency Consolidates

> They talk to me about progress, about 'achievements', diseases cured, improved standards of living. I am talking about societies drained of their essences, cultures trampled underfoot, institutions undermined, lands confiscated, religions smashed, magnificent artistic creations destroyed, extraordinary possibilities wiped out.
>
> —Aimé Césaire

This book has explored wind energy development in the Isthmus of Tehuantepec region in Oaxaca, Mexico, providing further insight from the front lines of wind energy development with the people fighting to defend their land, sea and cultural integrity. We learned that the project scale, placement, mitigation practices, and energy-use are foundational for assessing the viability and long-term socio-ecological sustainability of wind turbines. This research was conducted with a critical disposition towards anthropology. Anthropology and all social science research have the potential to be weaponized by both market and military forces and is openly and unapologetically on display. This is especially true in areas of natural resource extraction and conflict, where social scientists are employed to gather data and information on areas of social contestation as a means to pacify social tensions and undermine resistance movements (see Chapter 3). Such witting or unwitting academic collaboration and support undermines anthropological ethics. Anthropologists have near-endless institutional financial support if they work with police, military and resource extraction companies in conflict areas, using the low-intensity periods of a conflict zone—'peace times'—as an opportunity to monitor, take notes, collect data and perform the logistics of preemptive intelligence gathering on target populations. This was a problem in the Isthmus of Tehuantepec, where social scientists were used by or worked directly with wind companies to undermine resistance. This has heightened the local population's distrust of all researchers, who many times are viewed as the collaborators with wind companies and governmental forces.

The first chapter provided a brief history of agrarian politics and rebellion in the region, taking us from pre-colonial to super-colonial times with the rise of wind energy development. The second chapter addressed the activities in the northern coastal *Istmo* town of La Ventosa, the electrical infrastructure that engulfs the town with industrial wind turbines surrounding it. We learned who actually benefits from the wind turbine installations (*cacique*, the political authorities and some land owners, and, to a lesser extent, others, through temporary work, schools, roads and other civil infrastructure), as well as the irreversible damage caused to the people and their environment. Steeped in political corruption, land grabbing is conducted using leasing contracts and local middlemen employed by the wind companies to negotiate and/or acquire the land by any means necessary. The majority of people in the town, even some of those who were somewhat sympathetic or willing to work in wind energy, have become discontented, jaded or apathetic. '[E]ven if we are bothered', said one interview respondent, 'sometimes people cannot do anything and you know how politics works . . . the population does not decide what is going to be done and that is why we *conform to the things we are given*'. More importantly, there are serious and deadly health concerns that must be further investigated. People in La Ventosa complain about noise irritation, exhaustion, insomnia, headaches, dizziness and other symptoms related to what has been called the 'wind turbine syndrome'. Meanwhile, the large influx of cash into La Ventosa has created a kind of rural gentrification that has worsened the town's already existing poverty, crime, drug use and high rate of unemployment. Out-migration from La Ventosa appears to have increased as people seek work in other states in Mexico and the United States.

The Bíi Hioxo wind park in the southwest corner of Juchitán was the focus of Chapter 3. In Juchitán, the militant resistance that formed in opposition to the first (2014) wind park built on the Laguna Superior was undermined using textbook counterinsurgency strategies that mixed overt police, extra-judicial and legal repression with social development and public relations campaigns. This chapter described the tremendous impact of constructing a wind energy project that required the creation of social divisions and violent conflict. The destruction of the local ecology was facilitated by counterinsurgency warfare techniques, lightly armed police and heavily armed security personnel. This chapter revealed how sustainable development projects, like conventional energy, require coercive and violent repression against local populations to continue the trajectory of both capital accumulation and cultural and ecological transformation to a point of degradation and, potentially, destruction.

Chapter 4 ventured to the town of Álvaro Obregón, or Gui'Xhi' Ro, in Zapotec. In Álvaro Obregón, wind turbines were blocked from being constructed because of a determined opposition in and around the town. Here, as in Juchitán, it was clear to many residents that wind energy is instigating and reigniting old conflicts as well as creating new stresses on the natural environment. Wind energy development on the Santa Teresa Sand Bar would, and did, substantially impact the primarily indigenous populations living from the bounties of the land and sea in the area. The local government, land owners and *Ejidatarios*, hoping for prosperity, sold the town's collective future to Mareña Renovables, eventually sparking a popular insurrection for the land, sea and dignity of the town. The story of Álvaro Obregón is one of resistance, providing a glimpse into the reality of wind energy development and the complicated micro-politics of land acquisition, conflict and unrest in a semi-subsistence community. Wind energy in the culturally and ecologically sensitive Laguna Superior threatens to destructively alter the biosphere of the Santa Teresa sand bar. The uprising against the wind company, including battles with police and the takeover of the town hall, set the *cabildo comunitario* in conflict with the *constitucionalistas*. If cultural and biological preservation is a priority, industrial development ('green' or otherwise) requires a complete re-conceptualization in order to preserve harmony—socially and ecologically—with the land, and strengthen commitments to live *with* the cycles of the environment, not on top of it with degrading industrial interventions that further propel cycles of violence and communal discord.

Chapter 5 turned to the Free, Prior and Informed Consent (FPIC) consultation for the Eólica del Sur wind park. The consultation responded not only to the UN International Labor Organization's (ILO) Convention 169 that Mexico signed in 1990, but also to widespread uprisings against wind energy projects in the region. This exercise in democracy guided by the Mexican state sought to expand Indigenous rights and achieve greater incorporation into the wind project. Instead, however, it largely worked to legitimize the Eólica de Sur wind project's viability. The consultation failed to provide genuinely *free*, non-coercive information *prior* to approving the project, by default not allowing for *Informed Consent*. Instead, this process suffered from a direct conflict of interest between state and corporate collaboration; violated cultural norms; provided inadequate information; served as a marketing tool and platform instead of addressing the real issues of income, social and environmental impact concerns that had been experienced in the previous years. Consequently, state power became further entrenched within the existing political and economic asymmetry between actors who reinforced the illusions of real dialogue, negotiation and collective decision-making. The Mexican courts denied the Eólica del Sur after the FPIC consultation approved it, raising serious questions about the effectiveness of these consultations in affirming Indigenous decision-making and autonomy as well as questioning their operational purpose of legitimizing controversial development projects, which serves to pacify resistance. In this sense, the FPIC process can be seen as a de-escalation technique that benefits the state, regional elites and transnational development companies and further marginalizes the Zapotec, Ikoot and other fishermen and farmers seeking to retain their culture and protect the environments they derive sustenance from.

Chapter 6 took a more theoretical approach, arguing that wind energy in the *Istmo* is extending the processes of the industrial economy, grabbing Indigenous land and continuing the process of colonial genocide. I showed in Chapter 6 that industrial development and market processes are further entrenching the colonial model and ethos, resulting in a creeping 'slow industrial genocide' of cultural and biological diversity in the name of mitigating anthropogenic climate change (Huseman and

Short 2012). Finally, the chapter argued that the perception of wind energy as more benign than other methods of energy production such as oil, hydraulic fracturing, coal and nuclear energy, masks its dependence on fossil fuels and mineral extraction as well as its contribution to the continuation of ecological and cultural destruction.

With all the insights that have been offered up to this point, this book leads us to end with two final provocations for further consideration and research on renewable energy systems.

REBRANDING DYSTOPIA AND REBELLIOUS COMPLICITY

This book reveals how the destructive potential of wind energy is far greater than most people realize, especially for many of the people who position themselves as 'nature lovers', environmentalists, anti-capitalist, anti-industrialist or anti-state. While it should be recognized that the steel, copper, permanent magnets and electrical infrastructure of wind energy comes from the ground and thus from extractive processes (Dunlap 2018), this section offers two concerns regarding the renewable element in wind energy systems. The first is the potential of renewable energy, wind or otherwise, to help rebrand, renew and proliferate the dystopia of the present. The second is that people have been made complicit in the systemic forms of violence through the implicit intention of preserving the industrial economy, which manipulates their concerns for the environment and resolving climate crisis, leaving unquestioned the infrastructure that has been integrated into the everyday lives of people. It must be realized, by citizens and activists alike, that the destructive processes of colonialism, industrialism and the state are continued with wind energy development in its present form. Crucial to the colonial model are technological changes that promote separation, bureaucratization, industrial development and population control—in a word: civilization. For industrial infrastructure, systems of governance and capitalism to continue, technological development and renewable technologies are of high importance. All of this relies on the continuation and various forms of physical and cultural forms of genocide—the annihilation, assimilation, social death—or, at the least, pacification of anything that does not conform to market operations.

1. Rebranding Dystopia

Wind energy and other renewable technologies have the potential, and are on a trajectory, to rebrand dystopia. This means instead of creating utopias based on genuine ecological harmony, the current situation requires various forms of slavery, war and ecocide which have been cleverly disguised and rebranded and intensified. In order to understand what is at stake with renewable technologies under the present political, economic and ecological regime, an example from the past, specifically with corporate organizational evolution, will be used to demonstrate the technological shift slowly taking place with the rise of renewable energy systems.

Corporate ventures have been foundational to the development of nation-states, colonial conquest and economic development. Beginning with state-sanctioned charters, the East India, Hudson Bay and Unilever Company among others were born to spread across the world to accumulate, develop and expand the wealth of the nation (Nace 2003). Not only do these companies still exist, but so does this process of economic and industrial expansion. Corporate development and its organizational technology, however, also experienced severe limitations in the past, facing difficulties to manage the great logistical feat of corporate organizational expansion, profit and development. The technological limitations of corporate organization bear heuristic resemblance with energetic limitations of capitalist development to which renewable and wind energy are responding.

One corporate limitation worthy of mention emerged around World War I, which William Dugger (1988; 1989) locates as a key technological shift within the history of the corporation. This was the shift between the U-from and M-form corporate model (see figure C.1). The U-form model was divided into 'unitary' departments

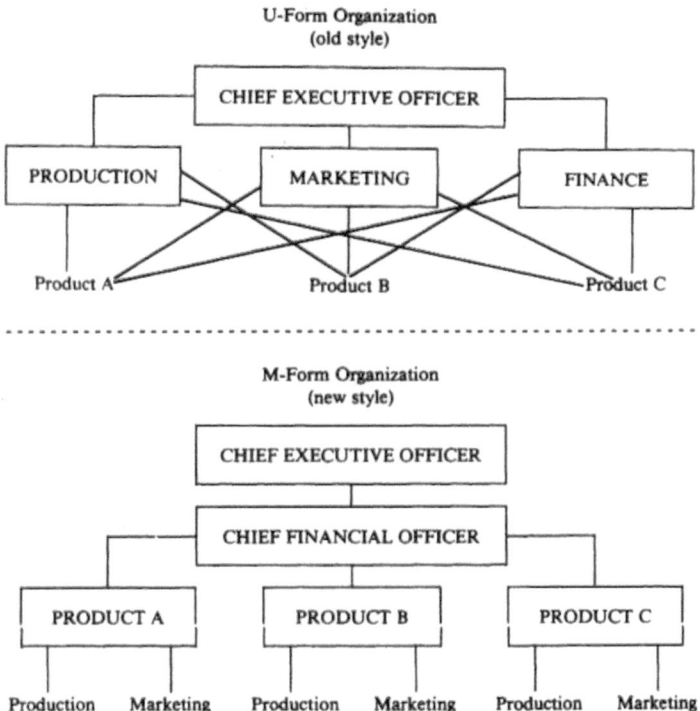

Figure C.1. The U-Form and M-Form corporate models.
SOURCE: DUGGER, 1988

to manage production as well as the organizational and economic growth of the corporation. This meant all the finance, production and marketing departments were grouped into their own administrative system to oversee corporate products (Dugger 1988; 1989). Production, marketing and finance departments were separated from each other, but had to work together to coordinate the production, marketing and finance of an array of corporate products. Eventually this led to inefficiencies, misunderstandings and the departments blaming each other for their (collective) failure of a particular product, which, as the company grew, resulted in significant profit losses. Related, and more importantly, the U-form structure prevented the realization (of the legally binding imperative) of profit maximization and shareholder value (Nace 2003). The corporation could not continue to expand and accumulate profit because of organizational/technological limitations, which needed to be overcome if the corporate entity would continue to grow.

Enter the 'M-form' model or 'multiple-divisions model' that first emerged in companies such a General Motors and Dupont after World War II (Dugger 1988; 1989). This model applied the approach of 'administrative decentralization', which retained its hierarchical function of the 'higher good' of corporate product development and shareholder value, but also organized the company into multiple divisions of more-or-less 'autonomous operating groups responsible for the production and distribution of a particular product or product line' (Dugger 1988, 82). The idea is that departments are formed around a particular product where marketing, production and finance staff worked closely together on the same product, as opposed to entire departments working separately on a plethora of different products. This reorganization encouraged group synergetic cooperation, created a high-level of product accountability, self-identification and responsibility for its success or failure, which further promoted greater employee integration, indoctrination and normalization of corporate culture (Dugger 1988; 1989). Dugger (1988, 82) writes: 'Each autonomous division performs its own operating functions, so that the head of each division can be held responsible for all of the functions contributing to the profitability of his group.'

The two important end results are first, profit responsibility is self-affirming and coded into the organizational technology itself and, second, the organizational possibility was created so the corporate organization could (theoretically) infinitely expand and grow. Whether the latter is possible in practice remains to be seen, yet the concentration and expansion of corporate power continues, and remains a tension driving the organizational life of different corporations today. The technological shift towards administrative decentralization is foundational to the spread of political economy and industrial infrastructure, not only in corporations, but also in the way that energy systems are designed and constructed. The shift from traditional fossil fuel to renewable energy systems is akin to a global U-to-M-form model shift, but on a global scale that is renewing the source of energy production and consequently the potential and organizational life of industrial infrastructure and capitalism itself. Wind energy in particular, and renewable energy in general, allows the state/corporate organizational form and its corresponding discursive and material infrastructures to spread further, possibility infinitely: accumulating and converting renewable resources to further industrial and capitalist development.

What the M-form model did for the corporate organization, renewable energy is now attempting to perform for neoliberal capitalism and its infrastructure. Renewable technologies are renewing and creating new markets around wind, solar, hydro and other natural resources, and with improvements in recycling, specifically in the sector of rare earth minerals and steel (see Habib and Wenzel 2014) that could allow for the infinite expansion of political economy and industrial infrastructure. Thus, renewable energy helps in slowly realizing the Christian project of total domination of the earth prescribed in Genesis 1:28: '. . . have dominion over the fish of the sea, and over the fowl of the air, and over every living thing that moveth upon the earth.' This is nothing short, as Perlman (2010/1983, 59) observed, of a 'declaration of war against' the earth, a war being advanced by not only grabbing the land and ecosystems where turbines are placed, but the vitality of the air that they harness and convert into the circuits of capitalist infrastructure.

The potential of infinite corporate growth, made possible by the M-form model, is the same potential realized with renewable energy systems and the growth possibilities being allotted to infrastructure development. In theory, renewable energy systems create the possibility of entrenching and spreading the industrial system across the entire planet. The political rationale of the 'lesser evil' examined by Eyal Weizman (2011, 10) is foundational to this merciless process of industrial development: the 'less brutal measures are also those that may be more easily naturalized, accepted and tolerated—and hence more frequently used, with the result that a greater evil may be reached cumulatively.'

Like panoptic architecture and Dugger's M-form model, both work to create self-managed and internalizing economical modes of governance and production: wind turbines not only intensify the industrial system but also sustain, continue and market the industrial economy to seem politically, ecologically and economically viable, in the face of intense ecological and social imbalances that are resulting in the ecological, climate and economic crises. The green economy is the lesser industrial evil, which works in the name of climate change mitigation in order to preserve the current system of industrialization that is dependent on spreading industrial waste and will do so with industrial-scale wind turbines. Stated simply, the green economy is the moral buffering of industrialism. The end result is greater cumulative social and environmental alteration, destruction, the cultural erasure of Indigenous people and even of counter-cultures, which are increasingly brought into the flows, institutions and business practices of the industrial economy. The fact remains: renewable energy systems strengthen capitalism, continue and create the possibility of spreading industrial infrastructure and the values it embodies across the entire world (Dalakoglou and Harvey 2012)—something fossil fuels are doing—but risk systemic suicide, social upheaval and insurrection that would necessitate change for this system to continue to exist. Capitalism is managing this change; meanwhile renewing its potential to usurp the vital energies of this planet—to energize its infrastructure and growth imperative.

All industrial-scale renewable energy infrastructures, even with high-levels of recycling, will necessitate enormous and increasing amounts of mineral extraction and energy to process and create renewable energy systems (see Dunlap 2018). Renewable energy, then, has the potential to make 'sustainable' the total extraction and reconfiguration of the earth along the lines of a capitalist utopia or human dystopia. This is why, for example, in science fiction films such as *Transcendence* (2014), or others concerning the emergence of sentient machines and artificial intelligence, the omnipresent machines take action early-on to build renewable energy facilities, such as a solar park in the desert or wind parks by the sea. Renewable energy (once recycling issues are sorted out) secures a continuous feed of energy to the sentient machine, thereby affirming the machine's autonomy or energy autonomy, allowing it to spread its hive-mind and infrastructure across the earth. This is to say, renewable energy is not only renewing destruction of landscapes and people, but it also holds the possibility to make capitalism and its infrastructure infinite and self-perpetuating with or without artificial intelligence or an omnipresent transhumanist monster. The 'sustainable' rhetoric of renewable energy masks the structural violence of capitalism, thereby renewing the destruction of techno-capitalist progress with little reflection or hesitation in the name of climate and ecological crises.

2. Rebellious Complicity

A person's dependence on, enchantment by, and enjoyment of the consumer products of the modern industrial economy creates a sort of Stockholm syndrome: one's identity becomes inextricably linked to the values and processes of this system, and one's life has been fused with it in such a way as to propel the logistical, economic and industrial organization (see Baudrillard 2010). Even so, many people remain discontented with the state of the world. One, among many, reasons for this discontent is global climate change, which has been receiving popular attention from national leaders and mainstream environmental social movements, notable among them are 350.org and groups such as Greenpeace and the Sierra Club. The movements protesting climate change tend to take a narrow view, focusing on carbon, greenhouse gas emissions and often uncritically supporting renewable energy systems. As I have shown, however, wind energy is renewing destruction and market growth while continuing the process of assimilating Indigenous populations. The shift discussed in Chapter 6 from classical colonialism to neocolonialism, entails a move from coercion to assimilation—or from obliteration to enslavement—as a seemingly 'friendlier' colonial/state system dependent on continued ecological destruction and various forms of slavery—agricultural planation, sweatshops, prison labour, mining and so on—to continue.

We are seeing a similar shift from traditional fossil fuels to renewable energy, as mentioned above. Positioning fossil fuels as the ultimate environmental destroyer then allows other forms of environmental destruction to remain hidden or politically acceptable. This is why industrial-scale renewable energy should be called 'fossil fuel+'. The + indicates the added benefit of the renewable component or multiplier present in renewable energy systems while simultaneously acknowledging their dependence on fossil fuel based-technologies and extractivism. The +, or renewable component, is dependent on fossil fuels in terms of the machines[1] used to mine the raw materials; transport those raw materials; power the processing facilities; transport the finished components; install and operate wind or other renewable energy systems (Dunlap 2018). The manufacturing assemblage of wind energy seems forgotten or glossed over, with a blind faith in life-cycle assessment modelling. Uncritical dispositions towards renewable energy systems paraded from corporate boardrooms to grass-roots social movements suffers a kind of reactive tendency—even a kind of Nietzschen ressentiment (see Newman, 2000)—that appears blind to the recuperative and flexible mechanism being developed by industry to prolong this trajectory of systemic profit-making and, consequently, industrial-scale destruction.

The environmentalists who condemn fossil fuels, nuclear and mineral extractivism are right, but this list must extend to acknowledge the reality of wind and other renewable energy systems. The narrow vision of ecological and social problems created by industrial development now combines with the uncritical acceptance of the new industrially marketed solutions of wind or other renewable energy systems. The end result is that the industrial system and the issues of climate change remains largely unaddressed as business booms and a green economy is born. Neglecting the realities of renewable energy implicitly accepts the existence of open-pit copper, iron ore and rare earth mining, which extends to other fossil fuel based industries and processes necessary to create renewable energy infrastructure/systems. Furthermore, there is a real risk that if fossil fuels are protested and successfully terminated in Europe or the Global North, then this type of extractivism will be out-sourced to other parts of Asia, Africa and Latin America.

In short, the colonial system—or civilization—needs to be brought into question as the solutions of politicians, corporations and NGOs to renew the destructive capacities of states and their economies once again (a classic of colonial discourse) forget about the extractive processes in the Global South: with fewer environmental regulations, increasing political violence and, more often than not, taking place within Indigenous territories. Politicians, corporations, NGOs and even some sections of environmental and climate movements are not radically reflecting on the personal, ecological and social problems that are generated by industrial development and its mode of production, especially when it comes to industrial-scale renewable energy systems. The demand for capitalist development is often used to justify and affirm this present trajectory, when in reality this self-affirming and fulfilling process needs to be broken. Critically reflecting on the manufacturing of socioeconomic demand and creating alternatives to development should remain a high priority. One of the greatest industrial degradations emerging from industrial development has been that of human imaginations, sensitivities and skills (see Illich

1973; 1978), which emerges from a political system necessitating that people turn against each other for not only survival, but also greater material and social status.

In a multifarious, uncoordinated and diverse approach, we should move away from the trajectory of techno-industrial progress globalized across the world. For example, articulating strategies of ecological self-sufficiency along the lines advocated by bioregionalism (Sale 1991)—but in our own individual and collective ways that defy every programme and suits the existing sociocultural context. This includes strategies of de-growth (D'Alisa 2014), post-development (Escobar 2015) and tearing up pavement so a pluriverse could breathe (Escobar 2018). Urban environments have endless potentials (Watkins 1993; Despommier 2010; Philips 2013; Hemenway 2015), but these require radical and concerted strategies to appropriate and rehabilitate ecosystems to live in harmony with the land. In this regard, local and Indigenous knowledge and its adaptation into urban permaculture techniques are insightful and inspiring. If there is a 'solution', it will depend on people's imaginations recreating their environments, reappropriating and transforming infrastructure to create, individually and collectively, healthy and sustainable environments.

This critique of wind and renewable energy and the environmentalists who support it serves to challenge everyone to imagine the alternatives that are stifled and destroyed by colonialism and our (often unwitting) participation in and dependence on this industrial system. This requires recognizing the systemic problems of political economy, affirming our individual-collective powers to engage our environments and experiment with small-scale energy systems as well as to engage with the infrastructure surrounding us to create ecologically, socially and organizationally sustainable environments. It also means challenging the mediated routines and relationships created by information technologies in our lives; appropriating these technologies as well as recognizing the influential power they exert over people; undoing the dichotomy of the individual and the collective by articulating strategies that value living with each other and our ecosystems; learning to combat forces of repression and recuperation by NGOs, municipal leaders and transnational wind energy corporations; questioning the dichotomy between fossil fuel and renewable energy. These tasks will not be easy. It depends on the will to navigate the seemingly impossible, to live life fully and to defend the values being eradicated and repackaged for sale by the techno-capitalist industrial system. Subverting ecological destruction, climate change and the authoritarianism that systematically degrades people and their eco-systems appears as an advisable course of action, if not a necessity.

NOTES

1. This is not to forget the fuels necessary for the factories and, overall, manufacturing assemblage necessary to create the mining equipment and machines used to acquire the raw materials for renewable energy systems.

Bibliography

Abram, D. *The Spell of the Sensuous,* New York: Vintage Books, 1996.
ADMEE. 'Panorama General de la Energía Eólica en México.' (2010): https://www.amdee.org/Amdee/AMDEE_presentacion_esp.pdf.
ADNsureste. 'En medio de protesta de la CNTE, Peña Nieto inaugura central eólica en Oaxaca.' (March 2016): http://www.24-horas.mx/maestros-de-la-seccion-22-marchan-en-oaxaca-contra-visita-de-pena-nieto/#.
Aguilar-Støen, M. 'Beyond transnational corporations, food and biofuels: the role of extractivism and agribusiness in land grabbing in Central America.' *Forum for Development Studies* 43, (2016): 155–75.
AI. (2014) 'Human Rights Challenges Facing Mexico: Amnesty International memorandum to President Enrique Peña Nieto.'(2014): https://www.amnesty.org/en/documents/AMR41/004/2014/en/.
AI. 'Amnesty International Report 2014/15: The State of the World's Human Rights.' (2015): https://www.amnesty.org/en/documents/pol10/0001/2015/en/.
Alexander, B. K. *The Globalization of Addiction: A Study in Poverty of the Spirit.* New York: Oxford University Press, 2008.
Altamirano-Jiménez, I. *Indigenous Encounters with Neoliberalism: Place, Women, and Environment in Canada and Mexico.* Vancouver: UBC Press, 2014.
Anaya, J. 'Observaciones del Profesor S. Jams Anaya Sobre la Consulta en el Contexto del Proyecto Energía Eólico del Sur en Juchitán de Zaragoza.' (2015): http://fundar.org.mx/wp-content/uploads/2015/03/Juchitan-observaciones-Anaya.pdf.
Anonymous. 'Accomplices Not Allies: Abolishing the Ally Industrial Complex.' (2014): http://www.indigenousaction.org/accomplices-not-allies-abolishing-the-ally-industrial-complex/.
Anthony, P. D. *The Ideology of War.* Oxon: Travistock Press/Routledge, 2001; 1977.
APIIDTT. 'Assembly of Indigenous Peoples of the Isthmus of Tehuantepec in Defense of Land and Territory.' *(APIIDTT) Blog.* (2012–2014): https://tierrayterritorio.wordpress.com/category/comunicados/english/.
APIIDTT. 'San Dionisio del Mar: Indigenous communities under police siege for resisting imposition of wind park.' (2013): http://elenemigocomun.net/2013/02/san-dionisio-del-mar-under-siege/.
APIIDTT. 'INCENDIAN ENRAMADA DE LA EX HACIENDA DEL GRAL. CHARIS.' (25 March 2013): https://tierrayterritorio.wordpress.com/category/videos/.

APIIDTT. 'Gunmen attack the assembly in Álvaro Obregón + Statement from the indigenous Peoples' Assembly for the Isthmus in Defense of the Land and Territory.' (2014):http://elenemigocomun.net/2014/03/gunmen-alvaro-obregon/.

APIIDTT. 'Communique: Presentation of observations on the Consultation in Juchitán to the former Rapporteur of the UN, James Anaya.' (Feb. 2015): Online: La Asamblea de Pueblos Indígenas del Istmo en Defensa de la Tierra y el Territorio.

APIIDTT. 'Comunicado urgente: Policía Municipal de Juchitán Ataca a Balazos a Compañeros de las Asambleas Comunitaria de *Álvaro* Obregón.' (May 2016): https://tierrayterritorio.wordpress.com/2016/05/15/comunicado-urgente-policia-municipal-de-juchitan-ataca-a-balazos-a-companeros-de-la-asamblea-comunitaria-de-alvaro-obregon/.

APIIDTT. 'Denunciamos el Asesinato del Compañero Rolando Crispín López a Manos de Policía Municipal de Juchitán de Zaragoza, Oaxaca.' (2018): https://tierrayterritorio.wordpress.com/2018/07/22/denunciamos-el-asesinato-del-companero-rolando-crispin-lopez-a-manos-de-policia-municipal-de-juchitan-de-zaragoza-oaxaca/.

Arendt, H. *The Origins of Totalitarianism.* New York: The World Publishing Company, 1962; 1951.

Arendt, H. *On Violence.* Orlando, FL: Houghton Mifflin Harcourt Publishing, 1970; 1969.

Argenti, N and D. M. Knight. 'Sun, wind, and the rebirth of extractive economies: renewable energy investment and metanarratives of crisis in Greece.' *Journal of the Royal Anthropological Institute* 21 (2015): 781–802.

Arronte, E. L., G. E. Castro-Soto, and T. P. Lewis. *Always Near, Always Far: The Armed Forces in Mexico,* edited by Ryan M. Zinn. San Francisco: Global Exchange, 2000.

Assies, W. 'Land Tenure and Tenure Regimes in Mexico: An Overview.' *Journal of Agrarian Change* 8 (2008): 33–63.

Avila, S. 'Environmental justice and the expanding geography of wind power conflicts.' *Sustainability Science* 13 (2018): 599–616.

Avila-Calero, S. 'Contesting energy transitions: wind power and conflicts in the Isthmus of Tehuantepec.' *Journal of Political Ecology* 24 (2017): 993–1012.

Baer, D. 'Here's Why Companies Are Desperate To Hire Anthropologists.'(2014): http://www.businessinsider.com/heres-why-companies-aredesperateto-hireanthropologists-2014-3?IR=T.

Baker, S. H. 'Why The IFC's Free, Prior, and Informed Consent Policy Does Not Matter (Yet) to Indigenous Communities Affected by Development Projects.' *Wisconsin International Law Journal* 30 (2013): 668–705.

Bakunin, M. 'God and the State.' In *No Masters, No God,* edited by D. Guérin, 151. Oakland: AK Press, 2005; 1871.

Bal, E., E. Grassiani and K. Kirk. 'Neoliberal individualism in Dutch universities: Teaching and learning anthropology in an insecure environment.' *Learning and Teaching* 7 (2014): 46–72.

Barelli, M. 'Free, prior and informed consent in the aftermath of the UN Declaration on the Rights of Indigenous Peoples: developments and challenges ahead.' *The International Journal of Human Rights* 16 (2012): 1–24.

Barnes, B. (2009) 'Disney Expert Uses Science to Draw Boy Viewers.' (2009): http://www.nytimes.com/2009/04/14/arts/television/14boys.html?em&_r=0.

Barta, T. 'Relations of genocide: land and lives in the colonization of Australia.' In *Genocide and the Modern Age,* edited by I. Wallimann and M.N. Dobkowski. Westport: Greenwood Press, 1987.

Bartra, A. 'Los municipios incómodos.' In *La remunicipalización de Chiapas: lo político y la politica en tiempo de contrainsurgencia*, edited by X.L. Solano and Mayor ABCy. México City: Cámara de Diputados LVII Legislatura, 2007.
Baudrillard, J. *The Agony of Power*. Los Angeles: Semiotext(e), 2010; 2005.
Bauer, Y. 'Whose Holocaust?' In *Genocide and Human Rights: A Global Anthology*, edited by J.N. Porter. Lanham, MD: University Press of America, 1982.
BBC. 'Mexico's Pena Nieto enacts major education reform.' (2013): http://www.bbc.co.uk/news/world-latin-america-21582629.
Bebbington, A. and J. Bury. *Subterranean struggles: New dynamics of mining, oil, and gas in Latin America*. Austin: University of Texas Press, 2013.
Behrens, P. and R. Henham. *Elements of Genocide*. Oxon: Routledge, 2013.
Bello, W. *Deglobalization*. New York: Palgrave Macmillan, 2004.
Bello, W. *The Food Wars*. New York: Verso Press, 2009.
Benjaminsen, T. A. and I. Bryceson. 'Conservation, green/blue grabbing and accumulation by dispossession in Tanzania.' *The Journal of Peasant Studies* 39 (2012): 335–55.
Bernays, E. 'The Engineering of Consent.' *The ANNALS of the American Academy of Political and Social Science* 250, no. 1(1947):113–20.
Bessi, R. and S. Navarro. 'Mexico: Researcher Raises Alert about Environmental Dangers of Wind Farms.' (2014a)://www.truth-out.org/news/item/26244-mexico-researcher-raises-alert-about-environmental-risks-in-region-with-highest-concentration-of-wind-farms-in-latin-america.
Bessi, R. and F. S. Navarro. 'Bíi Hioxo Wind Energy Project Hurting Indigenous Peoples and Their Territories.' (2014)://www.truth-out.org/news/item/23859-biio-hioxo-wind-energy-project-hurting-indigenous-peoples-and-their-territories.
Binford, L. 'Political Conflict and Land Tenure in the Mexican Isthmus of Tehuantepec.' *Journal of Latin American Studies* 17 (1985): 179–200.
Blakeley, R. *State terrorism and neoliberalism: The north in the south*. London: Routledge, 2009.
Blaser, Mario. 'Notes toward a political ontology of 'environmental' conflicts.' *Contested Ecologies: Dialogues in the South on Nature and Knowledge*, (2013):13–27.
Bnamericas. 'Bíi Hioxo Wind Farm.' (2015): http://www.bnamericas.com/project-profile/en/parque-eolico-bii-hioxo-parque-eolico-bii-hioxo.
Boas, H. *The No REDD Papers, Volume One*. Portland: Eberhardt Press, 2011: http://www.thecornerhouse.org.uk/sites/thecornerhouse.org.uk/files/No%20REDD%20papers%20One.pdf.
Böhm, S. and S. Dabhi. *Upsetting the Offset: The Political Economy of Carbon Markets*. London: MayFly Press, 2009.
Borras, S., C. Kay, S. Gómez, et al. 'Land grabbing and global capitalist accumulation: key features in Latin America.' *Canadian Journal of Development Studies /Revue canadienne d'e´tudes du de´veloppement* 33 (2012): 402–16.
Borras, S. M., P. McMichael and I. Scoones. 'The politics of biofuels, land and agrarian change: editors' introduction.' *Journal of Peasant Studies* 37 (2010): 575–92.
Bourdieu, P. *Acts of Resistance: Against the New Myths of Our Time*. Cambridge: Polity Press, 2000; 1998.
Bourgois, P. 'The Power of Violence in War and Peace: Post-Cold War Lessons from El Salvador.' *Ethnography* 2 (200): 5–34.
Bourgois, P. *In Search of Respect: Selling Crack in El Barrio*. Cambridge: Cambridge University Press, 2009; 1996.

Boyce, G. and C. Cash. 'Geography, Counterinsurgency, and the 'G-Bomb': The Case of Mécico Indígena.' In *Life During Wartime: Resisting Counterinsurgency*, edited by K. Williams, W. Munger, and L. Messersmith-Glavin. Edinburgh: AK Press, 2013.

Briseno, P. 'Frustran extorsión a empresa española; le pedían $500 mil.' (2016): http://www.excelsior.com.mx/nacional/2016/09/24/1118725.

Brock, A. Doctoral thesis: *Conserving Power: An exploration of biodiversity offsetting in Europe and beyond*. Brighton: University of Sussex Press, 2018.

Brock, A. and A. Dunlap. 'Normalising Corporate Counterinsurgency: engineering consent, managing resistance and greening destruction around the Hambach coal mine and beyond' *Political Geography* 62 (2018): 33–47.

Bryan, J. and D. Wood. *Weaponizing Maps: Indigenous peoples and Counterinsurgency in the Americas*. New York: The Guilford Press, 2015.

Bufe, C. and M. C. Verter. *Dreams of Freedom: A Ricardo Flores Magón Reader*. Oakland: AK Press, 2005.

Bull, B. and M. C. Aguilar-Støen. *Environmental Politics in Latin America: Elite dynamics, the left tide and sustainable development*. New York: Routledge Press, 2015.

Büscher, B., W. Dressler, and R. Fletcher. *Nature™ Inc: Environmental Conservation in the Neoliberal Age*. Tuscon: University of Arizona Press, 2014.

Büscher, B., S. Sullivan, K. Neves et al. 'Towards a synthesized critique of neoliberal conservation.' *Capitalism, Nature, Socialism* 23 (2012): 4–30.

Butt, D. 'Colonialism and postcolonialism.' *The International Encyclopedia of Ethics* (2013): 892–943.

Butzier, S. R. and S. M. Stevenson. 'Indigenous Peoples' Rights to Sacred Sites and Traditional Cultural Properties and the Role of Consultation and Free, Prior and Informed Consent.' *Journal of Energy and Natural Resources Law* 32 (2014): 297–334.

Calderón, Fernando Herrera, and Adela Cedillo, eds. *Challenging Authoritarianism in Mexico: Revolutionary Struggles and the Dirty War, 1964–1982*. New York: Routledge, 2012.

Call, W. *No Word for Welcome: The Mexican Village Faces the Global Economy*. Lincoln, NE: University of Nebraska Press, 2011.

Campbell, H. *Zapotec Renaissance: Ethnic Politics and Cultural Revivalism in Southern Mexico*., Albuquerque, NM: University of New Mexico Press, 1994.

Campbell, H., L. Binford, M. Bartolomé et al. *Zapotec Struggle: Histories, Politics, and Representations from Juchitán, Oaxaca*. Washington, DC: Smithsonian Institution Press, 1993.

Campbell, S. 'Mexican police brutally attack Oaxaca's striking teachers.' (June 2016): https://roarmag.org/essays/oaxaca-teacher-strike-police-attack/.

Çankaya, S. (2015) 'Professional Anomalies: Diversity policies policing ethnic minority police officers'. *European Journal of Policing Studies* 2, no. 4 (2015): 383–404.

Cariño, J. and M. Colchester. 'From Dams to Development Justice: Progress with "Free, Prior and Informed Consent" Since the World Commission on Dams.' *Water Alternatives* 3 (2010): 423–37.

Carlsen, L. 'Armoring NAFTA: The Battleground for Mexico's Future.' *North American Congress on Latin America (NACLA): Report on the Americas* 41 (2008): 17–20.

Castro-Gomez, S. 'The Social Sciences, Epistemic Violence, and the Problem of the 'Invention of the Other".' *Nepantal: Views from South* 3 (2002): 269–85.

CDM. '*CLEAN DEVELOPMENT MECHANISM: La Ventosa Wind Energy Project (the 'La Ventosa Project' or the 'Project').' Version: Document Version Number 3(2006)*: http://cdmloanscheme.org/sites/default/files/pdd_laventosa_0.pdf.

CDM. *Clean Development Mechanism Project Design Document Form (CDM-PDD): Fuerza y Energia Bii Hioxo Wind Farm*. (2012a): Online: United Nations Framework Convention on Climate Change (UNFCCC).

CDM. *CLEAN DEVELOPMENT MECHANISM: PROJECT DESIGN DOCUMENT FORM (CDM-PDD)-Version 03—in effect as of: 28 July 2006* (2012b): https://cdm.un fccc.int/filestorage/1/B/Z/1BZCF3LKGIUN4R2SATXV6Q7E8MJ9W0/San%20Dionisio%20Wind%20Farm%20-%20PDD%20resub.pdf?t=dWd8bzI0ZTN4fDB1unLi4PAZxayPaaebA-iq.

CencosTV. (20 Aug. 2015): http://www.ustream.tv/recorded/71255037

Césaire, A. *Discourses on Colonialism*. New York: Monthly Review Press, 1972; 1950.

Chaca, R. *El oro, la plata y la sal de los zoques, mixes y zapotecas*. (2015): http://www.noticias net.mx/portal/istmo/general/ambientales/275072-oro-plata-sal-zoques-mixes-zapotecas.

Chalk, F. 'Redefining Genocide.' In *Genocide: Conceptual and Historical Dimensions*, edited by G.J. Andreopolous. Philadelphia: University of Pennsylvania Press, 1994.

IIPoC. *IPCC Special Report on Renewable Energy Sources and Climate Change Mitigation Summary for Policymakers This Summary for Policymakers was formally approved at the 11th Session of Working Group III of the IPCC*. (2012): http://www.uncclearn.org/sites/default/files/inventory/ipcc15.pdf.

Chapman, S. 'The sickening truth about wind farm syndrome: Hilltop turbines are being blamed for myriad maladies. What is the truth behind these outlandish claims?' (2012): https://www.newscientist.com/article/mg21628850-200-the-sickening-truth-about-wind-farm-syndrome/.

Chatterjee, P. *The Nation and Its Fragments: Colonial and Postcolonial Histories*. Princeton: Princeton University Press, 1993.

Churchill, W. *Marxism and Native Americans*. Boston: South End Press, 1985.

Churchill, W. *A Little Matter of Genocide: Holocaust and Denial in the Americas 1492 to the Present*. San Francisco: City Lights Publisher, 2001.

Churchill, W. *Acts of Rebellion: The Ward Churchill Reader*. New York: Routledge Press, 2003.

Clastres, P. *Of Ethnocide. Archeology of Violence*. New York: Semiotext(e), 1994; 1980.

CODIGODH. *Rostros de la Impunidad en Oaxaca: Perspectivas desde la Defensa Integral de los Derechos Humanos,* Oaxaca: El Comité de Defensa Integral de Derechos Humanos Gobixha (CODIGO DH), 2014.

Cole T. 'The White-Savior Industrial Complex.' (2012): http://www.theatlantic.com/international/archive/2012/03/the-white-savior-industrial-complex/254843/.

CONEVAL. 'Informe de Evaluación de la Política de Desarrollo Social en México 2014.' (2015): http://www.coneval.org.mx/Informes/Evaluacion/IEPDS_2014/IEPDS_2014.pdf.

Consulta. 'Consulta Indígena en Juchitán.' (2015): https://consultaindigenajuchitan.word press.com/.

Copeland, N. 'Greening the Counterinsurgency: The Deceptive Effects of Guatemala's Rural Development Plan of 1970.' *Development and Change* 43 (2012): 975–98.

Copulsky, J. R. *Brand Resilience: Managing Risk and Recovery in a High-Speed World*. New York: Palgrave Macmillan, 2011.

Correa-Cabrera, G. *Los Zetas Inc. Criminal Corporations, Energy, and Civil War in Mexico*. Austin: University of Texas Press, 2017.

Corson, C., K. I. MacDonald, and B. Neimark, eds. 'Grabbing "Green": Markets, Environmental Governance and the Materialization of Natural Capital', Special Issue. *Human Geography* 6 (2013).

Costanza, J. N. 'Indigenous Peoples' Right to Prior Consultation: Transforming Human Rights from the Grassroots in Guatemala.' *Journal of Human Rights* 14 (2015): 260–85.

Craven, C. and D.A. Davis. *Feminist Activist Ethnography: Counterpoints to Neoliberalism in North America*. Lanham, MD: Lexington Books, 2013.

CrimethInc. 'From Democracy to Freedom.' (2016): http://crimethinc.com/texts/r/democracy/.

Crook, M. and D. Short. 'Marx, Lemkin and the genocide-ecocide nexus.' *The International Journal of Human Rights* 18 (2014): 298–319.

Curthoys, A. and J. Docker. 'Introduction—Genocide: definitions, questions, settler-colonies.' *Aboriginal History* 25 (2001): 1–15.

Cypher, J. M. 'Energy Privatized: The Ultimate Neoliberal Triumph.' *NACLA: Report on the Americas* 47 (2014): 32–31.

D'Alisa, G., F. Demaria, and G. Kallis. *Degrowth: a vocabulary for a new era*, London: Routledge, 2014.

Dalakoglou, D. and P. Harvey. 'Roads and Anthropology: Ethnographic Perspectives on Space, Time and (Im)mobility.' *Mobilities* 7 (2012): 459–65.

Danwatch. 'A People in the way of Progress: Prostitution, alcoholism and a lawsuit on illegal land acquisition in the Lake Turkana Wind Power project'. (2016): http://www.dr.dk/nyheder/htm/baggrund/generel/Danwatch.pdf.

Das, V. and D. Poole. *Anthropology in the Margins of the State*. Santa Fe, NM: School of American Research Press, 2004.

Davidson, M. and L. Lees. 'New-build "gentrification" and London's riverside renaissance.' *Environment and Planning A* 37(2005): 1165–190.

Davis, R. and M. Zannis. *The Genocide Machine in Canada*. Montreal: Black Rose Books, 1973.

Demarest, G. B. 'Property and Peace: Insurgency, Strategy and the Statute of Frauds'. (2009 [2007]): http://fmso.leavenworth.army.mil/documents/Property-and-Peace.pdf.

Denham, D. *Teaching rebellion: Stories from the grassroots mobilization in Oaxaca*. PM Press, 2008.

Despommier, D. *The Vertical Farm: Feeding the World in the 21st Century*, New York: St. Martin's Press, 2010.

Dowie, M. *Conservation Refugees: The Hundred-Year Conflict between Global Conservation and Native Peoples*, Cambridge, MA: MIT Press, 2009.

Downey, L., E. Bonds, and K. Clark. 'Natural Resource Extraction, Armed Violence, and Environmental Degradation'. *Organization Environment* 23(2010): 453–74.

Doyle, C. M. *Indigenous Peoples, Title to Territory, Rights and Resources: The Transformative Role of Free Prior and Informed Consent*, Oxon: Routledge, 2014.

Drost, P. N. *Genocide: United Nations Legislation on International Criminal Law*. Leiden: A.W. Sijthoff, 1959.

Duffy, R. *Killing for Conservation: The Politics of Wildlife in Zimbabwe*, Oxford: James Currey, 2000.

Duffy, R. 'War, by Conservation'. *Geoforum* 69 (2016): 238–48.

Dunlap, A. 'What is Progress? Revisiting Infrastructure and Civilizing Trends'. *Symposium on Technologies of Imperialism: Law in Contemporary and Historical Perspective*. University of London School of Oriental and African Studies (SOAS), 2013.

Dunlap, A. Power: Foucault, Dugger and Social Warfare. In *The BASTARD Chronicles: Social War*, edited by B. Collective, 55–106. Berkeley: Ardent Press, 2014a.

Dunlap, A. 'Permanent War: Grids, Boomerangs, and Counterinsurgency'. *Anarchist Studies* 22 (2014b): 55–79.

Dunlap, A. 'The Expanding Techniques of Progress: Agricultural Biotechnology and UN-REDD+'. *Review of Social Economy* 73 (2015a): 89–112.

Dunlap, A. 'The Coming Elections in Mexico: An Attack Against Indigenous Self-Determination in Álvaro Obregón'. (2015b): http://www.counterpunch.org/2015/06/04/the-coming-elections-in-mexico/.

Dunlap, A. 'Counter-insurgency: let's remember where prevention comes from and its implications'. *Critical Studies on Terrorism* 9 (2016).

Dunlap, A. 'The Genocide Machine Continues, Review of Redefining Genocide: Settler colonialism, social death and ecocide'. *Capitalism Nature Socialism* 28(2017): 118–20.

Dunlap, A. 'End the "Green" Delusions: Industrial-scale Renewable Energy is Fossil Fuel+'. (2018): https://www.versobooks.com/blogs/3797-end-the-green-delusions-industrial-scale-renewable-energy-is-fossil-fuel.

Dunlap, A "Agro sí, mina NO!" The Tía Maria Copper Mine, State Terrorism and Social War by Every Means in the Tambo Valley, Peru.' *Political Geography* 71, no. 1 (2019):10–25.

Dunlap, A. and J. Fairhead. 'The Militarisation and Marketisation of Nature: An Alternative Lens to "Climate-Conflict"'. *Geopolitics* 19 (2014): 937–61.

Dyer, Z. 'Clean Energy Plays Dirty in Oaxaca'. (2009): https://nacla.org/news/clean-energy-plays-dirty-oaxaca.

Earthfirst! 'Mexico: 22 Injured in Oaxaca Wind Farm Protest'. (2013): https://earthfirstnews.wordpress.com/2013/04/02/mexico-22-injured-in-oaxaca-wind-farm-protest/.

EB. 'Isthmus of Tehuantepec, Isthmus Mexico'. (2012): http://www.britannica.com/place/Isthmus-of-Tehuantepec.

EC. 'About the Equator Principles'. (2013): http://www.equator-principles.com/index.php/about-ep/38-about/about/352.

Edelman, M., C. Oya, and SMB Jr. 'Global Land Grabs: historical processes, theoretical and methodological implications and current trajectories.' *Third World Quarterly* 34 (2013): 1517–31.

EJOLT. 'Quilombola communities affected by wind power projects in Caetité region, Brazil'. (2016): https://ejatlas.org/conflict/quilombola-communities-affected-by-wind-power-projects-in-caetite-region-brazil.

EJOLT. 'Chiloé wind power project in Mapuche territory, Chile'. (2016a): https://ejatlas.org/conflict/chiloe-wind-power-project-in-mapuche-territory.

EJOLT. 'Chiloé wind power project in Mapuche territory, Chile'. (2016b): https://ejatlas.org/conflict/chiloe-wind-power-project-in-mapuche-territory.

Eldridge, E. R. and A. J. Reinke. 'Introduction: Ethnographic Engagement with Bureaucratic Violence'. *Conflict and Society* 4(2018): 94–8.

Elliott, D., M. Schwartz, G. Scott et al. *Wind Energy Resource Atlas of Oaxaca*. Golden, CO: National Renewable Energy Laboratory (NREL), 2003.

Ellul, J. *The Technological Society*, New York: Vintage Books, 1964; 1954.

Engle, K. *The Elusive Promise of Indigenous Development: Rights, Culture, Strategy*. Durham, NC: Duke University Press, 2010.

Erickson, C. L. 'Amazonia: the historical ecology of a domesticated landscape.' In *The Handbook of South American archaeology*, edited by H. Silverman and W. H. Isbell, 157–83. New York: Springer, 2008.

Escobar, A. 'Whose knowledge, whose nature? Biodiversity, conservation, and the political ecology of social movements'. *Journal of political ecology* 5 (1998): 53–82.

Escobar, A. *Encountering Development: the Making and Unmaking of the Third World.* Princeton: Princeton University Press, 2012; 1995.

Escobar, A., D. Rocheleau, and S. Kothari. 'Environmental social movements and the politics of place'. *Development* 45(2002): 28–36.

Escobar, A. 'Development, Violence and the New Imperial Order'. *Development* 47 (2004): 15–21.

Escobar, A. 'Degrowth, postdevelopment, and transitions: a preliminary conversation'. *Sustainability Science* 10 (2015): 451–62.

Escobar, A. *Designs for the pluriverse: radical interdependence, autonomy, and the making of worlds.* Durham, NC: Duke University Press, 2018.

Evans, A. 'Wind Farms and Health'. (2014): http://www.principia-scientific.org/wind-farms-and-health.html

Evans, R. and B. Thrope. 'The massacre of Aboriginal history'. *Overland* 163 (2001): 21–39.

Fairhead, J and M. Leach. *Misreading the African landscape: society and ecology in a forest-savanna mosaic.* London: Cambridge University Press, 1996.

Fairhead, J. and M. Leach. *Reframing deforestation: global analyses and local realities with studies in West Africa.* East Sussex: Psychology Press, 1998.

Fairhead, J. and M. Leach. *Science, Society and Power: Environmental Knowledge and Policy in West Africa and the Caribbean.* London: Cambridge University Press, 2003.

Fairhead, J., M. Leach, and I. Scoones. 'Green Grabbing: a new appropriation of Nature?' *Journal of Peasant Studies* 39 (2012): 237–61.

Fals-Borda, O. 'The Application of Participatory Action-Research in Latin America.' *International Sociology* 2(1987): 329–47.

Fanon, F. *The Wretched of the Earth.* New York: Grove Press, 1963; 1961.

Farbound, A., R. Crunkhorn, and A. Trinidade. '"Wind Turbine Syndrome": fact or fiction?' *The Journal of Laryngology and Otology* 127 (2013): 222–26.

Federici, S. *Caliban and the Witch: Women, The Body and primitive Accumulation.* New York: Autonomedia, 2009; 2004.

Ferguson, J. *The Anti-Politics Machine: Development, Depoliticization, and Bureaucratic Power in Lesotho.* Minnesota: University of Minnesota Press, 1994.

Feychting, M. 'Invited Commentary: Extremely Low-Frequency Magnetic Fields and Breast Cancer—Now It Is Enough!' *American Journal of Epidemiology* 178 (2013): 1046–50.

Flemmer, R. and A. Schilling-Vacaflor. 'Unfulfilled promises of the consultation approach: the limits to effective indigenous participation in Bolivia's and Peru's extractive industries'. *Third World Quarterly* 37 (2016): 172–88.

FM3-24. *Insurgencies and Countering Insurgencies.* (2014): http://fas.org/irp/doddir/army/fm3-24.pdf.

Fontana, L. B. and J. Grugel. 'The Politics of Indigenous Participation through "Free Prior and Informed Consent": Reflections from the Bolivian Case'. *World Development* 77 (2016): 249–61.

Foucault, M. *The Order of Things: Archaeology of the Human Sciences.* London: Tavistock Publications, 1977; 1966.

Foucault, M. The Eye of Power. In *Power/Knowledge: Selected Interviews & Other Writings, 1972–1977*, edited by C. Gordon, 146–65. New York: Pantheon Books, 1980; 1977.

Foucault, M. *Madness and Civilization: A History of Insanity.* New York: Random House Inc., 1989; 1961.
Foucault, M. *Discipline and Punish: The Birth of the Prison.* New York: Random House, 1995; 1977.
Foucault, M. 'Rituals of Exclusion'. In *Foucault Live: Collected Interviews, 1961–1984*, edited by S. Lotringer, 68–73. New York Semiotext(e), 1996; 1971.
Foucault, M. 'Confining Societies'. In *Foucault Live: Collected Interviews, 1961–1984*, edited by S. Lotringer, 83–94. New York Semiotext(e), 1996; 1972.
Foucault, M. 'Power Affects the Body'. In *Foucault Live: Collected Interviews, 1961–1984*, edited by S. Lotringer, 207–13. New York: Semiotext(e), 1996; 1977.
Foucault, M. 'Truth and Judicial Forms'. In *Power: Essential Works of Foucault, 1954–1984, Volume 3*, edited by J.D. Faubion, 1–89. London: Penguin, 2002; 1973.
Foucault, M. *'Society Must Be Defended:' Lectures at the College De France 1975–1976.* New York: Picador, 2003; 1997.
Foucault, M. The Meshes of Power. In *Space, Knowledge and Power: Foucault and Geography*, edited by J.W. Crampton and S. Elden. Farnham. UK: Ashgate Publishing Limited, 2007; 1976.
Foucault, M. *Security, Territory, Population: Lectures at the College De France 1977–1978.* New York: Picador, 2007; 1978.
Foucault, M. *The Birth of Biopolitics: Lectures at the College De France 1978–1979.* New York: Picador, 2008; 2004.
FPP (Forest Peoples Program). 'Making FPIC—Free, Prior and Informed Consent—Work: Challenges and Prospects for Indigenous Peoples'. (2007): http://www.forestpeoples.org/sites/fpp/files/publication/2010/08/fpicsynthesisjun07eng.pdf
Franco, J. 'Reclaiming Free Prior and Informed Consent (FPIC) in the context of global land grabs'. (2014): https://www.tni.org/files/download/reclaiming_fpic_0.pdf.
Friede, S. 'Enticed By the Wind: A Case Study in the Social and Historical Context of Wind Energy Development in Southern Mexico'. (2016): https://www.wilsoncenter.org/publication/enticed-the-wind-case-study-the-social-and-historical-context-wind-energy-development#sthash.rWabHFda.dpuf.
Gabrys, J. 'Programming environments: environmentality and citizen sensing in the smart city.' *Environment and Planning D: Society and Space* 32, no. 1 (2014):30–48.
Gadgil, M., F. Berkes, and C. Folke. 'Indigenous knowledge for biodiversity conservation'. *Ambio* (1993): 151–56.
Galeano, E. *Open Veins of Latin America: Five Centuries of the Pillage of a Continent.* London: Monthly Review Press, 1997; 1973.
Galeano, E. 'To Be Like Them'. In *The Post-Development Reader*, edited by M. Rahnema and V. Bawtree, 214–22. London: Zed Press, 1997; 1991.
Gamesa. 'Gamesa G90-2.0 MW'. (2008): http://www.wind-power-program.com/Library/Turbine%20leaflets/Gamesa/Gamesa%20G90%202mw.pdf.
Gedicks, A. 'Transnational mining corporations, the environment, and indigenous communities'. *The Brown Journal of World Affairs* 22 (2015): 129–52.
Gelderloos, P. *The Failure of Nonviolence: From Arab Spring to Occupy.* Seattle: Left Bank Books, 2013.
Gerber, J. F. and S. Veuthey. 'Plantations, Resistance and the Greening of the Agrarian Question in Coastal Ecuador'. *Journal of Agrarian Change* 10 (2010): 455–81.

Gibler, J. *Mexico Unconquered: Chronicles of Power and Revolt*. San Francisco: City Lights Books, 2009.
Giroux, H. A. *Neoliberalism's War on Higher Education*. Chicago: Haymarket Books, 2014.
Glaser, K. and S. T. Possony. *Victims of Politics: The State of Human Rights*. New York: Columbia University Press, 1979.
Gledhill, J. *The new war on the poor: the production of insecurity in Latin America*. London: Zed Books, 2015.
GNF-(Gas Natural Fenosa). '¿Quién es Bií Hioxo?/ Who is Bií Hioxo?' (2013a): https://biihioxo.wordpress.com/quien-es-bii-hioxo/.
GNF-(Gas Natural Fenosa). 'Mitos y realidades/Myth and Realities'. (2013b): https://biihioxo.wordpress.com/ecologia-y-cultura/.
GNF-(Gas Natural Fenosa). 'Se inauguró con éxito la muestra fotográfica 'Vientos de Cambio' del artista local Jacciel Morales/Successfully it inaugurated the photographic exhibition 'Winds of Change' by local artist Jacciel Morales'. (2013c): https://biihioxo.wordpress.com/2013/11/14/se-inauguro-con-exito-la-muestra-fotografica-vientos-de-cambio-del-artista-local-jacciel-morales/.
GNF-(Gas Natural Fenosa). 'Fuerza y Energía Bií Hioxo equipa a aulas de medios en Juchitán, a través del Programa ÚNETE/ Power and Energía *Bií Hioxo equips media classrooms in* Juchitán, through the JOIN program'. (2013d): https://biihioxo.wordpress.com/2013/11/28/fuerza-y-energia-bii-hioxo-equipa-a-aulas-de-medios-en-juchitan-a-traves-del-programa-unete/.
GNF-(Gas Natural Fenosa). 'Archivo de la etiqueta: Juchitán/ Tag Archives: BII Hioxo'. (2013e): https://biihioxo.wordpress.com/tag/juchitan/.
GNF-(Gas Natural Fenosa). '2014 Corporate Social Responsibility Report'. (2014): http://www.gasnaturalfenosa.com/servlet/ficheros/1297147982420/IRC_ing_accesible_op,0.pdf.
Gobierno. 'Condena Gobierno de Oaxaca hechos violentos registrados en Juchitán'. (2016): http://www.oaxaca.gob.mx/condena-gobierno-de-oaxaca-hechos-violentos-registrados-en-juchitan/.
González, J. A. A. 'The State Against the electricians: A War of Extinction'. *NACLA: Report on the Americas* 47 (2014): 56–9.
González, R. J. 'Taking the Next Step: Why We Should Continue Strengthening the AAA Ethics Code'. *Anthropology News* 50 (2009): 14–5.
González, R.J. *Militarizing Culture: Essays on the Warfare State*. Walnut Creek: Left Coast Press Inc., 2010.
González, R.J. 'The Rise and Fall of the Human Terrain System'. (2015): http://www.counterpunch.org/2015/06/29/the-rise-and-fall-of-the-human-terrain-system/.
González, R. J. 'Seeing into Hearts and Minds: Part 1. The Pentagon's Quest for a "Social Radar"'. *Anthropology Today* 31 (2015a): 8–13.
González, R. J. 'Seeing into Hearts and Minds: Part 2. 'Big data', Algorithms, and Computational Counterinsurgency'. *Anthropology Today* 31(2015b): 13–8.
Goodman, Barak, and Rachel Dretzin, producers. 'The Merchants of Cool'. Aired 27 February 2001 on *Frontline*, PBS.
Grajales, J. 'The rifle and the title: paramilitary violence, land grab and land control in Colombia'. *The Journal of Peasant Studies* 38 (2011): 771–92.
Grajales, J. 'State Involvement, Land Grabbing and Counter-Insurgency in Colombia'. *Development and Change* 44 (2013): 211–32.

Green, L. 'The Nobodies: Neoliberalism, Violence, and Migration'. *Medical Anthropology: Cross-Cultural Studies in Health and Illness* 30 (2011): 366–85.

Gross, T. 'Religion and Respeto: The Role and Value of Respect in Social Relations in Rural Oaxaca'. *Studies in World Christianity* 21 (2015): 119–39.

GrupoMexico. '2014 In Detail Annual Report'. (2014): http://www.gmexico.com/images/pdf/ReportesEng/Annual%20Report%202014%20Grupo%20M%C3%A9xico.pdf.

Guardian. 'Violence at Mexico teachers' protest leaves six dead, officials say'. (June 2016): https://www.theguardian.com/world/2016/jun/20/violence-mexico-teachers-protest-dead-oaxaca-union.

Guezuraga, B., R. Zaunera, and W. Pölz. 'Life cycle assessment of two different 2 MW class wind turbines'. *Renewable Energy* 37(2012): 37–44.

Gupta, A. and J. Ferguson. *Anthropological Locations: Boundaries and Grounds of a Field Science*. London: University of California Press, 1997.

Güven, F. *Decolonizing Democracy: Intersections of Philosophy and Postcolonial Theory*, London: Lexington Books, 2015.

Habib, K. and H. Wenzel. 'Exploring rare earth supply constraints for the emerging clean energy technologies and the role of recycling'. *Journal of Cleaner Production* 84 (2014): 348–59.

Hadjimihalis C. *Debt Crisis and Land Grabbing*. Athens: KΨM Publishers, 2014.

Hall, R., M. Edelman, S. M. Borras et al. 'Resistance, acquiescence or incorporation? An introduction to land grabbing and political reactions "from below"'. *Journal of Peasant Studies* 42 (2015): 467–88.

Hamister, L. 'Wind Development of Oaxaca, Mexico's Isthmus of Tehuantepec: Energy Efficient or Human Rights Deficient?' *Mexican Law Review* 5 (2012): 151–79.

Hannah, P. and F. Vanclay. 'Human rights, Indigenous peoples and the concept of Free, Prior and Informed Consent'. *Impact Assessment and Project Appraisal* 31(2013): 146–57.

Harrison, F. V. *Decolonizing Anthropology Moving Further Toward Anthropology for Liberation*. Washington DC: American Anthropological Association, 1991.

Havas, M. and D. Colling. 'Wind Turbines Make Waves: Why Some Residents Near Wind Turbines Become Ill'. *Bulletin of Science, Technology and Society* 31 (2011): 414–26.

Hemenway, T. *The Permaculture City: Regenerative Design for Urban, Suburban, and Town Resilience*. Vermont: Chelse Green Publishing, 2015.

Hildyard, N. *Licensed larceny: Infrastructure, financial extraction and the Global South*. Manchester: Manchester University Press, 2016.

Hinton, A. L. *Annihilating Difference: The Anthropology of Genocide*. Berkeley, CA: University of California Press, 2002.

Hinton, A. L. *Genocide: An Anthropological Reader*. Oxford: Blackwell Publishing, 2004.

Hoenderdaal, S., L. T. Espinoza, F. Marscheider-Weidemann et al. 'Can a dysprosium shortage threaten green energy technologies?' *Energy* 49 (2013): 344–55.

Hoffman, D. 'Frontline Anthropology: Research in a time of war'. *Anthropology Today* 19 (2003): 9–12.

Holmes, G. 'What is a land grab? Exploring green grabs, conservation, and private protected areas in southern Chile'. *Journal of Peasant Studies* 41(2014): 547–67.

Horkheimer, M. and T. W. Adorno. *Dialectic of Enlightenment' Philosophical Fragments*. Stanford, CA: Stanford University Press, 2002; 1944.

Howe, C. 'Anthropocenic Ecoauthority: The Winds of Oaxaca'. *Anthropological Quarterly* 87 (2014): 381–404.

Howe, C. and D. Boyer. 'Aeolian Politics'. *Distinktion: Scandinavian Journal of Social Theory* 16 (2015): 31–48.
Howe, C., D. Boyer, and E. Barrera. 'Wind at the Margins of the State: Autonomy and Renewable Energy Development in Southern Mexico'. In *Contested Powers*, edited by J.A. McNeish, A. Borchgrevink, and O. Logan. London: Zed Books, 2015.
HRW. 'Uniform Impunity: Mexico's Misuse of Military Justice to prosecute Abuses in Counternarcotics and Public Security Operations'. (2009): https://www.hrw.org/sites/default/files/reports/mexico0409web_0.pdf.
Huff, A. 'Black sands, green plans and vernacular (in) securities in the contested margins of south-western Madagascar'. *Peace building* 5 (2017): 153–69.
Huseman, J. and D. Short. '"A slow industrial genocide": tar sands and the indigenous peoples of northern Alberta'. *The International Journal of Human Rights* 16 (2012): 216–37.
Hyatt, S. B., B. W. Shear, and S. Wright. *Learning Under Neoliberalism: Ethnographies of governance in higher education*. Oxford: Berghahn, 2015.
IACHR. 'Indigenous and Tribal Peoples' Rights over their Ancestral Lands and Natural Resources: Norms and Jurisprudence of the Inter-American Human Rights System'. (2010): http://www.oas.org/en/iachr/indigenous/docs/pdf/AncestralLands.pdf.
Invisible Committee (IC). *To Our Friends*. South Pasadena, CA: Semiotext(e), 2015.
IDB. Marena Renovables Wind Power Project (ME-L1107) Environmental Category:A. *Environmental and Social Management Report (ESMR)* 15/6/2015.
IFC. 'International Finance Corporation's Guidance Notes: Performance Standards on Environmental and Social Sustainability'. (2012): http://www.ifc.org/wps/wcm/connect/e280ef804a0256609709ffd1a5d13d27/GN_English_2012_Full-Document.pdf?MOD=AJPERES.
IFC. 'Investments for a Windy Harvest: IFC Support of the Mexican Wind Sector Drives Results'. (2014): http://www.ifc.org/wps/wcm/connect/60c21580462e9c16983db99916182e35/IFC_CTF_Mexico.pdf?MOD=AJPERES.
Illich, I. *Energy and Equity*. London: Calder and Boyars, Ltd., 1973.
Illich, I. *Towards a History of Needs*. New York: Pantheon Books, 1978.
ILO. 'Indigenous and Tribal People's Rights in Practice: a guide to ILO Convention no. 169'. (2009): http://www.ilo.org/wcmsp5/groups/public/@ed_norm/@normes/documents/publication/wcms_106474.pdf.
Imparcial. 'EZLN y CNI piden autonomía y condenan violencia en Álvaro Obregón Piden que se respete el derecho del pueblo de elegir a sus a sus propias autoridades de manera autónoma'. (2016): http://imparcialoaxaca.mx/istmo/ap0/ezln-y-cni-piden-autonomía-y-condenan-violencia-en-álvaro-obregón.
IRENA. 'Renewable Power Generation Costs in 2014'. (2015): http://www.irena.org/documentdownloads/publications/irena_re_power_costs_2014_report.pdf.
Isenberg, A. C. *The Destruction of the Bison: An Environmental History. 1750–1920*. New York: Cambridge University Press, 2000.
Javers, E. 'Oil Executive: Military-Style 'Psy Ops' Experience Applied'. (2011): http://www.cnbc.com/id/45208498.
Jeffery, R. D., C. Krogh, B. Horner et al. 'Adverse health effects of industrial wind turbines'. *Canadian Family Physician /Le Médecin de famille canadien* 59 (2013): 921–25.
Jeffery, R. D., C. M. E. Krogh, and B. Horner 'Industrial wind turbines and adverse health effects'. *Can J Rural Med* 19 (2014): 21–26.

Jenss, A. 'Authoritarian neoliberal rescaling in Latin America: urban in/security and austerity in Oaxaca'. *Globalizations*, (2018) Latest Articles.
Johnson, T. M. and C. F. Sargent. *Medical Anthropology: Contemporary Theory and Method*. Westport: Praeger Publishers, 1996.
Jones, A. 'Gendercide and genocide'. *Journal of Genocide Research* 2(2000): 185–211.
Juárez-Hernández, S. and G. León. 'Wind Energy in the Isthmus of Tehuantepec: Development, Actors and Social Opposition'. *Problemas del Desarrollo: Revista Latinoamericana de Economía* 45 (2014): 1–9.
Juris, J. S. Practicing Militant Ethnography with the Movement for Global Resistance in Barcelona. In *Constituent Imagination: Militant Investigations: Militant Investigations // Collective Theorization*, edited by S. Shukaitis and D. Graeber D. Oakland, CA: AK Press, 2007.
Juris, J. S., and A. Khasnabish. *Insurgent Encounters: Transnational Activism, Ethnography, and the Political*. London: Duke University Press, 2013.
Kania, R. R. E. 'Joining Anthropology and Law Enforcement'. *Journal of Criminal Justice* 11(1983): 495–504.
Keith, D., J. DeCarolis, D. Denkenberger et al. 'The influence of large-scale wind power on global climate'. *Proc. Natl. Acad. Sci. USA* 101 (2004): 16115–120.
Keith, D.W., J.F. DeCarolis, D.C. Denkenberger et al. 'The influence of large-scale wind power on global climate'. *PNAS* 101 (2004): 16115–120.
Kilcullen, D. 'Twenty-Eight Articles: Fundamentals of Company-Level Counterinsurgency'. *IO Sphere* 2 (2006): 29–35.
Kingston, L. 'The Destruction of Identity: Cultural Genocide and Indigenous Peoples'. *Journal of Human Rights* 14 (2015): 63–83.
Kirsch, S. Sustainable Mining. *Dialectical Anthropology* 34 (2010): 87–93.
Kirsch, S. 'The social and the legal concept of genocide'. In *Elements of Genocide* edited by P. Behrens and R. Henham. Oxon: Routledge, 2013.
Kitson, F. *Low Intensity Operations: Subversion, Insurgency, and Peace Keeping*. London: Bloomsbury House, 2010; 1971.
Klein, N. *The Shock Doctrine: The Rise of Disaster Capitalism*, London: Allen Lane, 2007.
KM. 'México: Atacan a balazos a la comunidad Gui´xhi´ro´- Álvaro Obregón, Juchitán, Oaxaca'. (June 2015): http://kaosenlared.net/mexico-atacan-a-balazos-a-la-comunidad-guixhiro-alvaro-obregon-juchitan-oaxaca/.
Kohl, J. and J. Litt. *Urban Guerilla Warfare in Latin America*, Cambridge, MA: MIT Press, 1974.
Kothari, U. Power, Knowledge and Social Control in Participatory Development. In *Participation: The New Tyranny?* edited by B. Cooke and U. Kothari. London: Zed Books, 2001.
Kovats-Bernat, J. C. 'Negotiating Dangerous Fields: Pragmatic Strategies for Fieldwork amid Violence and Terror'. *American Anthropologist* 104(2002): 208–22.
Kuper, A. *The Reinvention of Primitive Society: Transformation of Myth*. Oxon: Routledge, 2005; 1988.
Kuper, A., K. Omura, E. Plaice et al. 'The return of the native'. *Current Anthropology* 44 (2003): 389–402.
Laurell, A. C. 'Three Decades of Neoliberalism in Mexico: The Destruction of Society'. *International Journal of Health Services* 45 (2014): 246–64.
Lawless, R. 'Empires and the Sullying of Anthropology'. (2009): http://www.counterpunch.org/2009/11/06/empires-and-the-sullying-of-anthropology/.

Lawrence, R. 'Internal colonisation and Indigenous resource sovereignty: wind power developments on traditional Saami lands'. *Environment and Planning D: Society and Space* 32 (2014): 1036–53.
Layton, R. *An Introduction to Theory in Anthropology*. Cambridge: Cambridge University Press, 1997.
Ledec, G. C., K. W. Rapp and R. G. Aiello. 'Greening the Wind: Environmental and Social Considerations for Wind Power Development in Latin America and Beyond'. (2011): http://www-wds.worldbank.org/external/default/WDSContentServer/WDSP/IB/2011/07/26/000333038_20110726003613/Rendered/PDF/634800v10WP0Gr00BOX361518B00PUBLIC0.pdf.
Lees, L., T. Slater, and E. Wyly. *Gentrification*. London: Routledge, 2008.
Lefort, C. *Democracy and Political Theory*. Minneapolis, MN: University of Minnesota Press, 1988.
Lemkin, R. *Axis rule in occupied Europe: laws of occupation, analysis of government, proposals for redress*. New Jersey: The Lawbook Exchange, 2005; 1944.
Levene, M. 'Empires, Native Peoples, and Genocide'. In *Empire, Colony, Genocide: Conquest, Occupation, and Subaltern Resistance in World History*, edited by A.D. Moses. Oxford: Berghahn, 2008.
Levene, M. and D. Conversi. 'Subsistence societies, globalisation, climate change and genocide: discourses of vulnerability and resilience'. *The International Journal of Human Rights* 18 (2014): 281–97.
Lewis, D. Anthropology and Colonialism. *Current Anthropology* 14 (1973): 581–602.
Li, T. M. 'Centering labor in the land grab debate'. *Journal of Peasant Studies* 38 (2011): 281–98.
Lloyd, D. and P. Wolfe. 'Settler colonial logics and the neoliberal regime'. *Settler Colonial Studies* 6 (2016): 109–18.
Lohmann, L. 'Carbon Trading: A Critical Conversation on Climate Change, Privatisation and Power'. *Development Dialogue* 48 (2006).
Lohmann, L. 'Carbon Trading, Climate Justice and the Production of Ignorance: Ten Examples'. *Development* 51 (2008): 359–65.
Lohmann, L. 'Climate as Investment'. *Development and Change* 40 (2009): 1063–83.
Lohmann, L. 'A Rejoinder to Matthew Paterson and Peter Newell'. *Development and Change* 45 (2012): 1177–84.
Lutz, C. 'The Military Normal'. In *The Counter-Counterinsurgency Manual: or, Notes on Demilitarizing American Society*, edited by Anthropologists NoC. Chicago: Prickly Paradigm Press, 2009.
Mahanty, S. and C. L. McDermott. 'How does "Free, Prior and Informed Consent" (FPIC) impact social equity? Lessons from mining and forestry and their implications for REDD+'. *Land Use* 35 (2013): 406–16.
Manzo, C. *Comunalidad, resistencia indígena y neocolonialismo en el Istmo de Tehuantepec, siglos XVI-XXI*. Mexico City: Ce-Acatl, 2011.
Manzo, D. 'Eólicas deben a Juchitán $800 millones en impuestos: alcalde'. (Jan. 2015): http://www.jornada.unam.mx/2015/01/27/estados/026n1est.
Manzo, D. 'Logran amparo para evitar construcción de parque eólico en Juchitán'. (Oct. 2015): http://pagina3.mx/2015/10/logran-amparo-para-evitar-construccion-de-parque-eolico-en-juchitan/.

Manzo, D. 'Teresita Luiz Ojeda, primera mujer ikotjs en ser alcaldesa en el Istmo de Tehuantepec'. (June 2016): http://www.istmopress.com.mx/especiales/teresita-luiz-ojeda-primera-mujer-ikotjs-alcaldesa-istmo-tehuantepec/?fb_ref=S6epKLibbu-Facebook.

Marijnen, E. and J. Verweijen. 'Selling green militarization: The discursive (re) production of militarized conservation in the Virunga National Park, Democratic Republic of the Congo'. *Geoforum* 75(2016): 274–85.

Marker, M. 'Indigenous voice, community, and epistemic violence: The ethnographer's "interests" and what "interests" the ethnographer'. *International Journal of Qualitative Studies in Education* 16 (2003): 361–75.

Markus, A. 'Genocide in Australia'. *Aboriginal History* 25 (2001): 50–70.

Márquez, P. O. 'Respuesta a la campaña de desprestigio de los funcionarios del Gobierno del Estado'. In *Self-published* edited by ADPSM. San Dionisio del Mar: Asamblea de Pueblos de San Dionisio del Mar (ADPSM), Fold-out brochure, 2013.

Marx, K. *Capital: A Critique of Political Economy Vol. I*. London: Penguin Books, 1982; 1867.

Mason, K. and P. Milbourne. 'Constructing a "landscape justice" for windfarm development: The case of Nant Y Moch, Wales'. *Geoforum* 53 (2014): 104–15.

Massé, F. and E. Lunstrum. 'Accumulation by securitization: Commercial poaching, neoliberal conservation, and the creation of new wildlife frontiers'. *Geoforum* 69 (2016): 227–37.

Mate, N. and S. Ghosh. 'The MSPL Wind Power CDM Project'. In *Upsetting the Offset: The Political Economy of Carbon Markets*. Edited by S. Böhm and S. Dabhi S. London: MayFly, 2009.

Matías, P. 'Un muerto y 440 incidentes enmarcaron elección en Oaxaca'. (June 2015): http://www.proceso.com.mx/406903/un-muerto-y-440-incidentes-enmarcaron-eleccion-en-oaxaca.

McAfee, K. and E. N. Shapiro. 'Payments for Ecosystem Services in Mexico: Nature, Neoliberalism, Social Movements, and the State'. *Annals of the Association of American Geographers* 100 (2010): 579–99.

McFate, M. 'Anthropology and Counterinsurgency: The Strange Story of their Curious Relationship'. *Military Review* 85 (2005): 24–38.

McGranahan, C., K. Roland, and B. C. Williams. 'Decolonizing Anthropology: A Conversation with Faye V. Harrison, Part I and II'. (2016): https://savageminds.org/2016/05/02/decolonizing-anthropology-a-conversation-with-faye-v-harrison-part-i/ & https://savageminds.org/2016/05/03/decolonizing-anthropology-a-conversation-with-faye-v-harrison-part-ii/.

Mejía, E. A. R. 'Mexico's wind energy and its energy economy'. (2014): http://www.renewablesinternational.net/mexicos-windenergy-and-its-energy-economy/150/435/82677/.

Mbembé, A. "Necropolitics." *Public Culture* 15, no. 1 (2003):11-40.

Members of the Network of Concerned A. 'The Network of Concerned Anthropologists Pledges to Boycott Counterinsurgency'. *Anthropology News* 48 (2007): 4–5.

Mendoza, L. 'Protestan 'anarcos' en Cuartel Militar y Penitenciaria de Oaxaca'. (Jan. 2015): http://www.noticiasnet.mx/portal/oaxaca/general/protestas/256867-protestan-anarcos-cuartel-militar-penitenciaria-oaxaca.

Menjívar, C. and N. Rodríguez. *When States Kill: Latin America, the U.S., and Technologies. of Terror*. Austin, TX: University of Texas Press, 2005.

Merchant, B. 'No, Wind Turbines Do Not Make Us Sick, Says Most Comprehensive Study Yet'. (2014): http://motherboard.vice.com/read/no-wind-turbines-do-not-make-us-sick-says-most-comprehensive-study-yet.

Merchant, C. *The Death of Nature: Women, Ecology, and The Scientific Revolution*. New York: Harper and Row, 1983.
Meyer, L. and B. M. Alvarado. *New World of Indigenous Resistance Open Media Series*. San Francisco: City Lights, 2010.
Meyers, D. T. *Victims' Stories and the Advancement of Human Rights*. New York: Oxford University Press, 2016.
MG (Mexican Government). 'Mexico's Constitution of 1917 with Amendments through 2007'. (2007): https://www.constituteproject.org/constitution/Mexico_2007.pdf.
Monjardin, A. L. 'Juchitán: Histories of Discord'. In: *Zapotec Struggle: Histories, Politics, and Representations from Juchitán, Oaxaca*, edited by H. Campbell, L. Binford, M. Bartolomé et al. Washington, DC: Smithsonian Institution Press, 1993.
Moses, A. D. 'Conceptual Blockages and definitional dilemmas in the 'racial century': genocides of indigenous peoples and the Holocaust'. *Patterns of Prejudice* 36 (2002): 7–36.
Moses, A. D. 'Empire, Colony, Genocide: Conquest, Occupation, and Subaltern Resistance in World History' in *War and Genocide*. Oxford: Berghahn, 2010; 2008.
MR (Mareña Renovables). 'Mareña Renovables statement'. (2014): http://business-human rights.org/sites/default/files/media/documents/company_responses/marena_renovables _re_wind_farm_mexico.pdf.
Munday, M., G. Bristow, and R. Cowell. 'Wind farms in rural areas: How far do community benefits form wind farms represent a local economic development opportunity?' *Journal of Rural Studies* 27 (2011): 1–12.
Murphy, J. 'Place and exile: resource conflicts and sustainability in Gaelic Ireland and Scotland'. *Local Environment* 18 (2013): 801–16.
NAACP-LDF. 'Media Highlights in Police Shooting of Michael Brown'. (n.d.): http://www .naacpldf.org/press-release/media-highlights-police-shooting-michael-brown.
Nace, T. *Gangs of America: The Rise of Corporate Power and the Disabling of Democracy*. San Francisco: Berrett-Koehler Publishers, Inc., 2003.
NACLA. 'Mexico: The State Against the Working Class'. *NACLA (North American Congress on Latin America): Report on the Americas* 47 (2014).
Nahmad, S. 'El Impacto Social del Uso del Recurso Eólico'. (2011): https://langleruben.files .wordpress.com/2014/06/1-informe-final-ec3b3lico.pdf.
Nauman, T. 'Oaxaca's wind farm surge produces clean power—and protests'. (2014): http:// www.trust.org/item/20140605124619-k9pav/.
Navarro, F. S. 'Energy Reform, Shale Gas and Public Spending Cuts in Mexico'. (2013): http://www.truth-out.org/news/item/19078-energy-reform-shale-gas-and-public-spending -cuts-in-mexico.
Newman, S. 'Anarchism and the Politics of Ressentiment.' *Theory and Event* 4, no. 3: (2000) 3–34. https://theanarchistlibrary.org/library/saul-newman-anarchism-and-the-politics-of -ressentiment.
Nocella, A. J., S. Best, and P. McLaren. *Academic Repression: Reflections from the Academic Industrial Complex*. Oakland, CA: AK Press, 2010.
Nordstrom, C. and A. C. G. M. Robben. *FieldWork Under Fire: Contemporary Studies of Violence and Survival*. London: University of California Press, 1997.
Norget, K. 'Caught in the Crossfire: Militarization, Paramilitarization, and State Violence in Oaxaca, Mexico'. In *When States Kill: Latin America, the U.S., and Technologies of Terror*, edited by C. Menjívar and N. Rodríguez. Austin, TX: University of Texas Press, 2005.

Nuijten, M. *Power, Community and the State: The Political Anthropology of Organization in Mexico.* London: Pluto Press, 2003.

Oceransky, S. 'Fighting the Enclosure of Wind: Indigenous Resistance to the Privatization of the Wind Resource in Southern Mexico'. In *Sparking a Worldwide Energy Revolution: Social Struggles in the Transition to a Post-Petrol World,* edited by K. Abramsky. Oakland, CA: AK Press, 2011.

Osborne, T. 'Fixing Carbon, Losing Ground: Payments For Environmental Services and Land (In)Security in Mexico'. *Human Geography* 6 (2013): 119–33.

Overbeek, W., M. Kröger, and J. F. Gerber. 'An Overview of Industrial Tree Plantations in the Global South: Conflicts, Trends and Challenges for Resistance Struggles'. In *EJOLT (Environmental Justice-Organization-Liabilities and Trade),* edited by L. Lohmann. Barcelona: Autonomous University of Barcelona, 2012.

Owens, P. *Economy of Force: Counterinsurgency and the Historical Rise of the Social.* Cambridge: Cambridge University Press, 2015.

Padel, F. and S. Das. 'Cultural genocide and the rhetoric of sustainable mining in East India'. *Contemporary South Asia* 18 (2010): 333–41.

Paley, D. *Drug War Capitalism.* Oakland, CA: AK Press, 2014.

Paley D. 'Ayotzinapa: Paradigm of the War on Drugs in Mexico. New Afterword to Drug War Capitalism'. (2015): http://www.revolutionbythebook.akpress.org/new-afterword-to-dawn-paleys-drug-war-capitalism/.

Palmer, A. *Colonial Genocide.* Adelaide: Crawford House, 2000.

Pasqualetti, M. 'Wind energy landscapes: society and technology'. *Society and Natural Resources* 14 (2001): 689–99.

Pasqualetti, M. J. 'Mobility, Space and the Power of Wind-Energy Landscapes'. *Geographical Review* 90: (2000) 381–94.

Pasqualetti, M. J. 'Opposing Wind Energy Landscapes: A Search for Common Cause'. *Annals of the Association of American Geographers* 101 (2011): 907–17.

Payan, T., and G. Correa-Cabrera. *Issue Brief 10.29.14: Land Ownership and Use Under Mexico's Energy Reform.* (2014): https://bakerinstitute.org/files/8400/.

Peluso, N. and C. Lund. 'New Frontiers of Land Control: Introduction'. *The Journal of Peasant Studies* 38 (2011): 667–81.

Peluso, N. L. 'Coercing Conservation? The politics of state resource control'. *Global Environmental Change* 3 (1993): 199–217.

Peluso, N. L. and P. Vandergeest. 'Political Ecologies of War and Forests: Counterinsurgencies and the Making of National Natures'. *Annals of the Association of American Geographers* 101 (2011): 587–608.

Peñoles. *Our Great People: 2014 Annual Report.* (2014): http://www.penoles.com.mx/wPortal/content/conn/UCM/path/Carpetas/www/English/Press%20Room/Annual%20Reports/PENOLES-2014-ANNUAL-REPORT.pdf;jsessionid=fz8hVLPFb6P2pnSY4BgMnfpvG7xtsSY11Q71cqvhP8JbLpn8lC32!1983562992.

Perera, J. 'Old Win in New Bottles: Self-determination, Participatory Democracy and Free, Prior and Informed Consent'. In *Indigenous Studies and Engaged Anthropology: The Collavorative Moment,* edited by P. Sillitoe. Farnham: Ashgate, 2015.

Perlman, F. *The Continuing Appeal of Nationalism.* Detroit: Red and Black Press, 1985.

Perlman, F. *Against His-story, Against Leviathan* Detroit: Red and Black Press, 1983

Petersen, R. 'Mexico: Federal Court Halts Controversial Wind Park'.(Dec. 2012): https://globalvoices.org/2012/12/27/mexico-federal-court-halts-controversial-wind-park/.

Phadke, R. 'Resisting and Reconciling Big Wind: Middle landscape Politics in the New American West'. *Antipode* 43 (2011): 754–76.

Philips, A. *Designing Urban Agriculture: A Complete Guide to the Planning, Design, Construction, Maintenance and Management of Edible Landscapes*. London: Blackwell-Wiley Publishing, 2013.

Phillips, M. 'Rural Gentrification and the Processes of Class Colonisation'. *Journal of Rural Studies* 9 (1993): 123–40.

Pierpont, N. *Wind Turbine Syndrome: A Report on a Natural Experiment*. Lowell: K-Selected Books, 2009.

Piggott, H. *A Wind Turbine Recipe Book*: Metric Edition, 2009.

Pike, F. B. 'The municipality and the system of checks and balances in Spanish American colonial administration'. *The Americas* 15 (1958): 139–58.

Pizarro, V. '"I've had enough", says Mexican attorney general in missing students gaffe.' (2014): http://www.theguardian.com/world/2014/nov/09/protests-flare-in-mexico-after-attorney-generals-enough-im-tired-remarks.

Platt, J. 'Why Winona LaDuke is fighting for food sovereignty'. (2013): http://www.mnn.com/leaderboard/stories/why-winona-laduke-is-fighting-for-food-sovereignty.

PODER. 'Pese a existir 32 incidentes de seguridad y 75 solicitudes de información sin atender, la Consulta en Juchitán pasa a su Fase Deliberativa'. (2015): http://projectpoder.org/es/2015/06/1982/.

Polanyi, M. *Personal Knowledge: Towards a Post-Critical Philosophy*. Chicago: University of Chicago Press, 1974; 1958.

Powell, C. 'What do genocides kill? A relational conception of genocide'. *Journal of Genocide Research* 9 (2007): 527–47.

Price, D. H. *Weaponizing Anthropology: Social Science in Service of the Militarized State*. Oakland, CA: AK Press/Counterpunch Books, 2011.

Price, D. H. 'Counterinsurgency by Other Names: Complicating Humanitarian Applied Anthropology in Current, Former, and Future War Zones'. *Human Organization* 73 (2014): 95–105.

Quigley J. 'Genocide: A Useful Legal Category?' *International Criminal Justice Review* 19 (2009): 115–31.

Rabinow, P. 'Comments'. *Current Anthropology* 36 (1995): 430–33.

Rahnema, M. and V. Bawtree. *The Post-Development Reader*. London: Zed Books, 1997.

Rambo, A. T. 'Primitive Polluters. Semang Impact on the Malaysian Tropical Rain Forest Ecosystem'. *Anthropological Papers* (1985): 1–98.

Raymond, H. 'The ecologically noble savage debate. *Annu. Rev. Anthropol.* 36 (2007): 177–90.

Reuters. 'Mexican government says poverty rate rose to 46.2 percent in 2014'. (2015): http://www.reuters.com/article/us-mexico-poverty-idUSKCN0PX2B320150723.

RF. 'Return Fire'. (2013-Present): https://325.nostate.net/distro/.

Ribot, J. and N. Peluso. 'A theory of access'. *Rural Sociology* 68 (2003): 153–81.

Rivas, S. C. 'Consulta Definirá Futuro De Inversiones Eólicas: Ocaso o resplandor'. (Jan. 2015): http://www.noticiasnet.mx/portal/sites/default/files/flipping_book/oax/2015/01/23/secc_a/files/assets/basic-html/page20.html.

Robins, S. 'On the Call for a Militant Anthropology: The Complexity of "Doing the Right Thing"'. *Current Anthropology* 37 (1996): 341–43.

Robinson, W. I. '(Mal)Development in Central America: Globalization and Social Change'. *Development and Change* 29 (1998): 467–97.
Robson, J. and F. Berkes. 'How Does Out-Migration Affect Community Institutions? A Study of Two Indigenous Municipalities in Oaxaca, Mexico'. *Human Ecology* 39 (2011): 179–90.
Robson, J. P. 'Local approaches to biodiversity conservation: lessons from Oaxaca, southern Mexico'. *International Journal of Sustainable Development* 10 (2007): 267–86.
Rocheleau, D. E. 'Networked, rooted and territorial: green grabbing and resistance in Chiapas'. *The Journal of Peasant Studies* 42 (2015): 695–723.
Rodgers, D. and B. O'Neill. 'Infrastructural Violence: Introduction to the Special Issue'. *Ethnography* 13 (2012): 401–12.
Rodríguez-Garavito, C. 'Ethnicity.gov: Global Governance, Indigenous Peoples and the Right to Prior Consultation in Social minefields'. *Indiana Journal of Global Legal Studies* 18 (2011): 263–305.
Rostow, W. *The Stages of Economic Growth: A Non-Communist Manifesto*. Cambridge: Cambridge University Press, 1960.
Roth, J. P. *The Logistics of the Roman Army at War: 264 B.C.–A.D. 235*. Boston: Brill, 1999.
Rotz, S. 'REDD'ing Forest Conservation: The Philippine Predicament'. *Capitalism, Nature, Socialism* 25 (2014): 43–59.
RS. 'Management Discussion and analysis for second quarter ended March 31, 2009'. (2008): http://rivres.com/images/financial_statements/2009-Q1-MDA.pdf.
Rubí, M. 'Arranca segunda fase de Central Eólica Sureste I'. (March 2016): http://eleconomista.com.mx/estados/2016/03/03/arranca-segunda-fase-central-eolica-sureste-i.
Rubin, J. W. *Decentering the Regime: Ethnicity, Radicalism, and Democracy in Juchitán, Mexico*. Durham, NC: Duke University Press, 1997.
Sahlins, M. *How 'Natives' Think: About Captain Cook For Example*. Chicago: University of Chicago Press, 1996.
Said, E. *The Question of Palestine*. New York: Vintage Books, 1996; 1979.
Sale, K. *Dwellers in the Lands: Bioregional Vison*. Athens, GA: Georgia University Press, 1991; 1985.
Salemink, O. 'Social science intervention: Moral Versus political economy and the Vietnam War'. In *A Moral Critique of Development: In search of global Responsibilities*, edited by P.Q.V. Ufford and A.K. Giri. London: Routledge, 2003.
Salman, T. and W. Assies. 'Anthropology and the Study of Social Movements'. In *Handbook of Social Movements Across Disciplines*, edited by B. Klandermans and C. Roggeband, 205–65. New York: Springer, 2007.
Sarte, J-P. 'Genocide'. *New Left Review* 48 (1968): 13–25.
Schaller, D. J. 'From Conquest to Genocide: Colonial Rule in German Southwest Africa and German East Africa'. In *Empire, Colony, Genocide: Conquest, Occupation, and Subaltern Resistance in World History*, edited by A.D. Moses. Oxford: Berghahn, 2008.
Scheper-Hughes, N. *Death Without Weeping: The violence of Everyday Life in Northeast Brazil*. Berkeley, CA: University of California Press, 1992.
Scheper-Hughes, N. 'The Primacy of the Ethical: Propositions for a Militant Anthropology'. *Current Anthropology* 36 (1995): 409–40.
Scheper-Hughes, N. and P. Bourgois. *Violence in War and Peace: An Anthology*. Oxford: Blackwell Publishing, 2004.
Schutter, O. D. 'The Green Rush: The Global Race for Farmland and the Rights of Land Users'. *Harvard International Law Journal* 52 (2011): 503–59.

Scoones, I, R. Hall, S. M. Borras et al. 'The politics of evidence: methodologies for understanding the global land rush'. *Journal of Peasant Studies* 40 (2013): 469–83.
Scott, J. C. *Seeing Like a State: How Certain Schemes to Improve the Human Condition Have Failed.* New Haven, CT: Yale University Press, 1998.
Seagle, C. 'Inverting the Impacts: Mining, Conservation and Sustainability Claims near the Rio Tinto/Qmm Ilmenite Mine in Southeast Madagascar'. *Journal of Peasant Studies* 39 (2012): 447–77.
SEDESOL. 'Catálogo de Localidades'. (2010): http://www.microrregiones.gob.mx/catloc/LocdeMun.aspx?tipo=clave&campo=loc&ent=20&mun=043.
See, S. E. 'Accumulating the primitive'. *Settler Colonial Studies* 6 (2015): 164–73.
Sharma, A. and A. Gupta. *The anthropology of the state: a reader.* London: John Wiley and Sons, 2006.
Shaw, M. *What is Genocide?* Cambridge: Polity Press, 2007.
Short, D. 'Cultural Genocide and indigenous peoples: A sociological approach'. *The International Journal of Human Rights* 14 (2010): 833–48.
Shpiro-Garza, E. 'Contesting Market-Based Conservation: Payments for Ecosystem Services as a Surface of Engagement for Rural Social Movements in Mexico'. *Human Geography* 6 (2013): 119–33.
Shukaitis, S. and D. Graeber. *Constituent Imagination: Militant Investigations: Militant Investigations // Collective Theorization.* Oakland, CA: AK Press, 2007.
Siamanta, Z. C. *Greening energy production in post-crisis Greece: a political ecology analysis of economic crisis, 'the green economy' and the neoliberalisation of nature,* PhD dissertation. London: Birkbeck University of London, 2016.
Siamanta, Z. C. 'Building a green economy of low carbon: the Greek post-crisis experience of photovoltaics and financial "green grabbing"'. *Journal of Political Ecology* 24 (2017): 258–76.
Sierra, M. T. 'The Revival of Indigenous Justice in Mexico: Challenges for Human Rights and the State'. *PoLAR: Political and Legal Anthropology Review* 28 (2005): 52–72.
Simon, S. 'Friction in a Warming World: The Challenges of Green Energy in Rural Oaxaca, Mexico'. *Journal of Peace, Conflict and Development* 20 (2013): 6–19.
Simpson, J. 'Do Police Departments Need Anthropologists?' (2014): https://anthropoliteia.net/2014/12/08/do-police-departments-need-anthropologists/.
Sin-Embargo. 'Juez ordena suspender parque eólico en Juchitán, Oaxaca; es un logro de pueblos indígenas: ONGs'. (Dec. 2015): http://www.sinembargo.mx/16-12-2015/1578990.
SIPAZ. 'In Focus: Impacts and effects of the wind-energy projects in the Tehuantepec Isthmus'. (2013): http://www.sipaz.org/en/reports/117-informe-sipaz-vol-xviii-no-3-septiembre-de-2013/468-enfoque-impactos-y-afectaciones-de-los-proyectos-de-energia-eolica-en-el-istmo-de-tehuantepec.html.
SIPAZ. 'Oaxaca: Conflict over extraordinary elections in San Dionisio del Mar'. (Dec. 2104): https://sipazen.wordpress.com/2014/12/30/oaxaca-conflicts-over-extraordinary-elections-in-san-dionisio-del-mar/.
SIPAZ-Blog. 'SIPAZ Blog: The International Service for Peace Blog'. (2012–2015): https://sipazen.wordpress.com/.
Sklair, L. 'Social Movements for Global Capitalism: The Transnational Capitalist Class in Action'. *Review of International Political Economy* 4(1997): 514–38.
Sklair, L. (2001) *The Transnational Capitalist Class.* Oxford: Blackwell Publishers, 2001.

Smith, B. T. *Pistoleros and Popular Movements: The Politics of State Formation in Postrevolutionary Oaxaca.* Lincoln, NE: University of Nebraska Press, 2009.

Smith, D. P. 'Extending the Temporal and Spatial Limits of Gentrification: A Research Agenda for Population Geographers'. *International Journal of Population Geography* 8 (2002): 385–94.

Smith, D. P. 'What is Rural Gentrification? Exclusionary Migration, Population Change, and Revalorised Housing Market'. *Interface: Planning Theory and Practice* 12 (2011): 596–605.

Smith, J. M. 'Indigenous Communities in Mexico Fight Corporate Wind Farms'. (2012): http://upsidedownworld.org/main/mexico-archives-79/3952-indigenous-communities-in-mexico-fight-corporate-wind-farms.

Smith, R. W. 'Human destructiveness and politics: the twentieth century as an age of genocide'. In *Genocide and the Modern Age,* edited by I. Wallimann and M.N. Dobkowski. Westport: Greenwood Press, 1987.

Solovey, M. 'Project Camelot and the 1960s Epistemological Revolution: Rethinking the Politics-Patronage-Social Science Nexus'. *Social Studies of Science* 31(2001): 171–206.

Southern, P. *The Roman Army: A Social and Institutional History.* Oxford: ABC-CLIO, 2006.

Sparks, C. *Globalization, Development and the Mass Media.* London: SAGE Publications Ltd., 2007.

Spivak, G. 'Can the Subaltern Speak?' In *Marxism and the Interpretation of Culture,* edited by C. Nelson and L. Grossberg. Urbana: University of Illinois Press, 1988, 271–313.

Springer, S. *The discourse of neoliberalism: An anatomy of a powerful idea.* London: Rowman and Littlefield International, 2016.

Starzmann, M. T. 'Anti-Anthropology'. *Social Anthropology* 24 (2016): 375–76.

Stephen, L. 'The Construction of Indigenous Suspects: Militarizaton and the gendered and ethnic dynamics of human rights abuses in Southern Mexico'. *American Ethnologist* 26 (2000): 822–42.

Stephen, L. *Zapata Lives! Histories and Cultural Politics in Southern Mexico.* Berkeley: University of California Press, 2002.

Stephen, L. 'Negotiating Global, National, and Local "Rights" in a Zapotec Community'. *PoLAR: Political and Legal Anthropology Review* 28 (2005): 133–50.

Stephen L. *Transborder Lives: Indigenous Oaxacans in Mexico, California, and Oregon.* Durham, NC: Duke University Press, 2007.

Stephen, L. *We are the face of Oaxaca: Testimony and Social Movements.* Durham, NC: Duke University Press, 2013.

Sullivan, S. 'Elephant in the room? Problematising "new"(neoliberal) biodiversity conservation'. *Forum for Development Studies* 33, no. 1 (2006):105–35.

Sullivan, S. 'Green Capitalism, and the Cultural Poverty of Constructing Nature as Service Provider'. *Radical Anthropology* 3 (2009): 18–27.

Sullivan, S. '"Ecosystem service commodities"—a new imperial ecology? Implications for animist immanent ecologies, with Deleuze and Guattari'. *New Formations: A Journal of Culture/Theory/Politics* 69 (2010): 111–28.

Sullivan, S. 'On Bioculturalism, Shamanism and Unlearning the Creed of Growth'. *Geography and You,* March 2011, 15–19.

Sullivan, S. 'Banking Nature? The Spectacular Financialisation of Environmental Conservation'. *Antipode* 45 (2013a).

Sullivan, S. 'After the Green Rush? Biodiversity Offsets, Uranium Power and The "Calculus of Casualties" in Greening Growth'. *Human Geography* 6 (2013b): 80–101.

Sullivan, S. 'The natural capital myth; or will accounting save the world?' *LCSV Working Paper Series No. 3*. The University of Manchester: The Leverhulme Centre for the Study of Value: School of Environment, Education and Development, (2014): 1–42.

Sullivan, S. 'On "Natural Capital", "Fairy Tales" and Ideology.' *Development and Change* 48.2 (2017a): 397–423.

Sullivan, S. 'What's ontology got to do with it? On nature and knowledge in a political ecology of the "green economy"'. *Journal of Political Ecology* 24 (2017b): 217–42.

Tabassum-Abbasi, M. Premalatha, T. Abbasi et al. 'Wind energy: Increasing deployment, rising environmental concerns'. *Renewable and Sustainable Energy Reviews* 31 (2014): 270–88.

Taussing, M. *Shamanism, Colonialism and the Wild Man: A Study in Terror and Healing*. Chicago: University of Chicago Press, 1987.

The Counted. 'The Counted: People killed by police in the US'. (n.d.): https://www.theguardian.com/us-news/ng-interactive/2015/jun/01/the-counted-police-killings-us-database.

Thompson L. 'Roman Roads'. *History Today* 47 (1997): 21–28.

TRC. 'Honouring the Truth, Reconciling for the Future: Summary of the Final Report of the Truth and Reconciliation Commission of Canada'. (2015): http://www.trc.ca/websites/trcinstitution/File/2015/Findings/Exec_Summary_2015_05_31_web_o.pdf.

Truman, H. 'Inaugural Address'. (1949): http://www.presidency.ucsb.edu/ws/?pid=13282.

Tucker, K. *For Wildness and Anarchy*. Canada: Black and Green Press, 2010.

Tutino, J. 'Ethnic Resistance: Juchitán in Mexico History'. In *Zapotec Struggle: Histories, Politics, and Representations from Juchitán, Oaxaca*, edited by H. Campbell, L. Binford, M. Bartolomé et al. Washington, DC: Smithsonian Institution Press, 1993.

Ulloa, A. *The Ecological Native: Indigenous Peoples' Movements and Eco-Governmentality in Columbia*. London: Routledge, 2013; 2005.

UNDRIP. *United Nations Declaration on the Rights of Indigenous Peoples*. (2008): http://www.un.org/esa/socdev/unpfii/documents/DRIPS_en.pdf.

UNFCCC. 'Sustainable Energy for All: A Global Action Agenda, Pathways for Concerted Action toward Sustainable Energy for All'. (2012): http://www.un.org/wcm/webdav/site/sustainableenergyforall/shared/Documents/SEFA-Action%20Agenda-Final.pdf.

UNGC. 'What is UN Global Compact?' (2014): https://www.unglobalcompact.org/what-is-gc.

USAID. 'Mexico Wind Farm Case Study'. (2009a): http://www.energytoolbox.org/gcre/wind_case_study.pdf.

USAID. 'Elementos para la Promoción de la Energía Eólica en México'. (2009b): https://energypedia.info/wiki/File:Elementos_para_la_Promoci%C3%B3n_de_la_Energ%C3%ADa_E%C3%B3lica_en_M%C3%A9xico.pdf.

Vance, E. 'The Wind rush: Green Energy blows Trouble into Mexico'. (2012): http://www.csmonitor.com/Environment/2012/0126/The-wind-rush-Green-energy-blows-trouble-into-Mexico.

Vásquez, V. R. M. 'Autoritarismo, movimiento popular y crisis política: Oaxaca'. Oaxaca: Instituto de Investigaciones sociológicas, Universidad Autónoma 'Benito Juárez' de Oaxaca, 2007.

Vazquez, D. G. 'In Defense of the Territory of Life: A Look Into the Territory of the Community Police in Guerrero, Mexico'. In *Why Don't the Poor Rise up?* edited by A. Nangwaya and M. Truscello, 213–26. Oakland, CA: AK Press, 2017.

Veblen, T. *The Higher Learning In America: A Memorandum on the Conduct of Universities By Business Men*. New York: Sentry Press, 1965; 1918.

Veblen, T. *Theory of the Leisure Class*. New York: Oxford University Press, 2009; 1899.
Veracini, L. 'Understanding colonialism and settler colonialism as distinct formations.' *Interventions* 16, no. 5 (2014):615–33.
Verweijen, J. and Marijnen, E. 'The counterinsurgency/conservation nexus: guerrilla livelihoods and the dynamics of conflict and violence in the Virunga National Park, Democratic Republic of the Congo'. *The Journal of Peasant Studies* 45 (2018 [2016]): 300–20.
Vidal, J. 'The Great Green Land Grab'. (2008): http://www.theguardian.com/environment/2008/feb/13/conservation.
Virilio, P. *The Art of the Motor*. Minneapolis: University of Minnesota Press, 1995.
Virilio, P. *Pure War*. Los Angeles: Semiotext(e), 2008; 1983.
Virilio, P. *War and Cinema: The Logistics of Perception*. New York: Verso Press, 2009; 1984.
Wang, C. and R. G. Prinn. 'Potential climatic impacts and reliability of very large-scale wind farms'. *Atmos Chem Phys* 10(2010): 2053–61.
Warman, A. The Future of the Isthmus and the Juárez Dam. In *Zapotec Struggle: Histories, Politics, and Representations from Juchitán, Oaxaca*, edited by H. Campbell, L. Binford, M. Bartolomé et al. Washington, DC: Smithsonian Institution Press, 1993.
Watkins, D. *Urban Permaculture: A Practical Handbook for Sustainable Living*. London: Permanent Publications, 1993.
WDM. 'Power to the People? How World Bank Financed Wind Farms Fail Communities in Mexico'. (2011): http://www.ritimo.org/IMG/pdf/Power_to_the_people_.pdf.
Weizman, E. *The Least of All Possible Evils: Humanitarian Violence from Arendt to Gaza*. London: Verso Press, 2011.
White, B., S. M. Borras, R. Hall et al. 'The new enclosures: critical perspectives on corporate land deals'. *Journal of Peasant Studies* 39 (2012): 619–47.
WHO. 'Night Noise Guidelines for Europe'. (2009): http://www.euro.who.int/__data/assets/pdf_file/0017/43316/E92845.pdf.
Wijk, R. V., J. V. D. Greef, and E. V. Wijk. 'Human Ultraweak Photon Emission and the Yin Yang Concept of Chinese Medicine'. *Journal of Acupuncture Medicine* 3 (2010): 221–31.
Williams, K. *Our Enemies in Blue: Police and Power in America*. Cambridge: South End Press, 2007; 2004.
Williams, K. 'The other side of the COIN: Counterinsurgency and community policing'. *Interface* 3 (2011): 81–117.
Williams, K., W. Munger, and L. Messersmith-Glavin. *Life During Wartime: Resisting Counterinsurgency*. Edinburgh: AK Press, 2013.
Wilson J. 'The Urbanization of the Countryside: Depoliticization and the Production of Space in Chiapas'. *Latin American Perspectives* 189 (2013): 2.
Wilson, J. 'Model Villages in the neoliberal era: the Millennium Development Goals and the colonization of everyday life'. *Journal of Peasant Studies* 41(2014): 107–25.
Windustry. 'How much do wind turbines cost?' (2012): http://www.windustry.org/how_much_do_wind_turbines_cost.
Wise, R. D. and H. M. Covarrubias. 'Capitalist Restructuring, Development and Labour Migration: the Mexico–US case'. *Third World Quarterly* 29 (2008): 1359–74.
Wolfe, P. *Settler Colonialism and the Transformation of Anthropology*. London: Routledge Press, 1999.
Wolfe, P. 'Settler colonialism and the elimination of the native'. *Journal of Genocide Research* 8 (2006): 387–409.

Wolfe, P. 'Structure and Event: Settler Colonialism, Time, and the Question of Genocide'. In *Empire, Colony Genocide: Conquest, Occupation, and Subaltern Resistance in World History*, edited by A. D. Moses, 102–32. New York: Berghahn Books, 2010; 2008.

Wolfe, P. 'After the Frontier: Separation and Absorption in US Indian Policy'. *Settler Colonial Studies* 1 (2011): 13–51.

Wolford, W., S. M. Borras, R. Hall et al. 'Governing Global Land Deals: The Role of the State in the Rush for Land'. *Development and Change* 44 (2013): 189–210.

Wolsink, M. 'Entanglement of interests and motives: assumptions behind the NIMBY theory on facility siting'. *Urban Studies* 31 (1994): 851–66.

Wolsink, M. 'Wind Power and the NIMBY-myth: institutional capacity and the limited significance of public support'. *Renewable Energy* 21 (2000): 49–64.

Wolsink, M. 'Planning of renewable schemes: deliberative and fair decision-making on landscape issues instead of reproachful accusations of non-cooperation'. *Energy Policy* 35 (2007): 2692–704.

Woolford, A. 'Discipline, Territory, and Colonial Mesh: Indigenous Boarding Schools in the United States and Canada'. In *Colonial Genocide in Indigenous North America*, edited by A. Woolford, J. Benvenuto, and A.L. Hinton. Durham, NC: Duke University Press, 2014.

Woolford, A., J. Benvenuto, and A. L. Hinton. *Colonial Genocide in Indigenous North America*. Durham, NC: Duke University Press, 2014.

X A. 'Give Up Activism'. (2009): https://theanarchistlibrary.org/library/andrew-x-give-up-activism.

Yannakakis, Y. *The Art of Being In-between: Native intermediaries, Indian identity, and local rule in colonial Oaxaca*. Durham, NC: Duke University Press, 2008.

Ybarra, M. 'Taming the Jungle, Saving the Maya Forest: Sedimented Counterinsurgency Practices in Contemporary Guatemalan Conservation'. *The Journal of Peasant Studies* 39 (2012): 479–502.

Zeitlin, J. F. 'Ranchers and Indians on the Southern Isthmus of Tehuantepec: Economic Change and Indigenous Survival in Colonial Mexico'. *The Hispanic American Historical Review* 69 (1989): 23–60.

Zeitlin, J.F. *Cultural Politics in Colonial Tehuantepec Community and State among the Isthmus Zapotec, 1500–1750*. Stanford, CA: Stanford University Press, 2005.

Zografos, C. and J. Martínez-Alier. 'The politics of landscape value: a case study of wind farm conflict in rural Catalonia'. *Environment and Planning: A* 41 (2009): 1726–44.

Zoomers, A. 'Globalisation and the foreignisation of space: seven processes driving the current global land grab'. *Journal of Peasant Studies* 37(2010): 429–47.

Index

Acciona Energía, 44
acculturation, 32
activism, environmental, 20
adaptation, colonial incursion, resistance and, 29–33
ADPSM. *See* People's Assembly of San Dionisio del Mar
AEI. *See* State Agency of Investigations
agrarian politics, 54–55
agriculture, 70, 142, 149
Alliance for Prosperity and Peace, 53
Altamirano-Jiménez, Isabel, 31
Álvaro Obregón (Gui'Xhi' Ro), xxiv, 6–7, 21–23, 103; blood feuds in, 122; COCEI support in, 107–8; communal land in, 107–8; Communitarians seizure of town hall of, 114–15; elections in, 115–17, 120; fisherman in, 108; gazebo mural in, 8; history of, 105; resistance in, 9; town square mural, *123*; violence in, 122; wind energy companies in, 107–9
anarchists, 6–7, 151
Anaya, James, 22
animal mortality, wind turbines and, 62–64, 96–97, 163
anthropogenic climate change, 77
anthropologists, 3, 9–10, 13–15
anthropology, 9, 12, 14, 15–16, 169
anti-anthropology, 10
APIIDTT. *See* Assembly of Indigenous Peoples of the Isthmus of Tehuantepec in Defense of Land and Territory

APIITDTT. *See* Assembly of the Isthmus in Defense of the Land and Territory
APPJ. *See* Popular Assembly of the Juchiteco People
APPO. *See* Popular Assembly of the Peoples of Oaxaca
Aquino, Zárate, 38
arrests, 110
artificial intelligence, 176
Asamblea Popular de los Pueblos de Oaxaca (Popular Assembly of the Peoples of Oaxaca) (APPO), 43
Assembly of Indigenous Peoples of the Isthmus of Tehuantepec in Defense of Land and Territory (APIIDTT), 134–35
Assembly of the Isthmus in Defense of the Land and Territory (APIITDTT), xxvi
Association of Indigenous Communities in the Northern Zone of the Isthmus (UCIZONI), 60
Attica, 151
authoritarianism, 7, 178
autoabastecimiento ('self-supply'), 44, 50
autonomy, 34
Axis Rule in Occupied Europe (Lemkin), 152–53
Ayotzinapa massacre, xxi, xxiii, 4
Ayuntamiento Popular (people's government), 39

Barra de Santa Teresa (*Barra*), 2, 6, 24–25, 110–16, 128, 157, 170

203

Battle for the Barra, 113
benefit sharing, unequal, 73n5
Benito Juárez Dam, 24, 36–37
betrayal, 27n8
Big Bear, 119, 121, 162
Bíi Hioxo wind park, 1, 22, 44, 170; APPJ and, 75–76; communal land and, 83–84; construction of, 24, 75, 83, 87–88; DFIG of, 99; human rights and, 95; investors in, 77; land leasing of, 84–85; monetary shaping operations of, 93; as operational, 96–98; planning for, 77; public relations campaign, 94–95, 95; reforestation project, 93; resistance and protests against, 75–76, 78, 87; security cameras at, 87; social development programs of, 92–94; wind turbines of, 77, 83–88
Binford, Leigh, 37
biodiversity, 97
birds, 62, 97, 98, 119, 157, 160, 163, 165, 167n11
blood feuds, 122
Borras, Saturnino, 18–19
Brazil, 18
bribery, 107
BRIC countries (Brazil, Russia, India and China), 18
Bustillo, Mario, 38

cabildo comunitario. *See* community counsel
cacique. *See* political boss
caciquismo, insurrection and, 33–37
capitalism, 3, 26, 99–100, 121, 154, 164, 172, 174–75
caravanas de la muerte (Caravans of Death), 43
carbon dioxide reduction, 164
Cárdenas, Lázaro, 34, 105–6
cartel violence, 73n8
Catholic Church, 31, 120
CDHT. *See* Tepeyac Center for Human Rights
CDM. *See* clean development mechanism
Central American Plan Pueblo Panama, 42
Central Light and Power (LyFC), xxii
Central State Penitentiary Ixcotel, xxviii

certified emissions reduction credits (CERs), 52
Césaire, Aimé, 169
CFE. *See* Federal Electricity Commission
Charis Castro, Heliodoro, 24, 36–37, 47n4, 105–7, 111
Chatterjee, Partha, 36
Chile, 18
Chontals, 29
Christianity, 31–32
CIFs. *See* Climate Investment Funds
civil conflict, wind energy development and, 113, 138
civil dissent, 79, 80
clean development mechanism (CDM), 52, 72, 77, 84, 99, 165
Clean Technology Fund (CTF), 52
climate change, xxiii, 2, 20, 72, 77, 99–100, 104, 123, 149, 156–57, 177
Climate Investment Funds (CIFs), 52
coal, 20, 25, 98, 149, 172
COCEI. *See* Isthmus Coalition of Workers, Peasants, and Students
coercion, 96, 98, 104, 155
Cold War, 79
collective land commissioner (*comisariado*), 108, 110
colonial genocide, 148, 152–56
colonial incursion, adaptation, resistance and, 29–33
colonialism/colonization, 14, 16, 25, 100, 129, 152, 172, 176, 178; adaptation to, 33; agriculture and, 149; characteristics of, 150; core of, 149; Güven on, 151; having viral and bacterial qualities, 150; imperialism and, 149; model, 149–51; modernization and, 154; Moses on, 150; settler colonialism and, 150, 160
colonial law, 145
comisariado. *See* collective land commissioner
Comisión Federal de Electricidad. *See* Federal Electricity Commission
communal land (*ejidos*), xxii, 34–35, 45, 170–71; in Álvaro Obregón, 107–8; Bíi Hioxo wind park and, 83–84; conflict over, 37; credits for, 38; freedom and,

110; of Isthmus of Tehuantepec, 106; legal status of Juchitán, 83; Mareña Renovables wind park and, 110; Mexican constitution and, 80; privatization of, 85; sacred sites and, 86
Communal Land Commission, 106
Communitarian Police (*polícia comunitaria*), xxvi, 7–9, 23, *111*, *113*, 114–20
community armed groups, 125n11
community counsel (*cabildo comunitario*), 7, 9, 105, 118–19, 171
community development, 90
compañerxs, 90, 112, 119
comunalidad education, 4
CONAFOR. *See* Mexican National Forestry Commission
concertación accords (*Pacto de Concertación Social*), 41
conflict management, 77
conquest, neoliberalism and, 154–55
conservation, 19, 82
constitucionalistas (Constitutionalists), 105
constitution: Mexican, 80, 107, 115, 125n5; Oaxacan, 114
Constitutionalists (*Los Contras*), 8–9, 115–20, 126n21, 171
consumer products, 176
consumption, 99
contaminated water, 64
continuity, 35
contracts, 138
Los Contras. *See* Constitutionalists
corn, 27n9, 85
corporate development, 173
corporate power, *173*
corporate social responsibility (CSR), 95, 130, 133, 163
corruption, 6, 40, 55–56, 60, 67, 120, 136, 146n5
Cortés, Juan, Don, 31
cost of living, 142–43
counter-information tactics, 93
counterinsurgency, 20–21; aid, 89; employment and, 82–83; FPIC and, 145; greening of, 76–77, 82; hard/direct, 78, 88–91, 96; hot spots, 91–96; Mexican history and, 126n16; in

Mexico, 79–83; military and, 77; poverty and, 79; property and, 81, 82; regional strategies, 82; soft/indirect, 78, 91, 96; strategies, 170; sustainability and, 82; tactics as genocide, 155; techniques for, 24, 78, 98, 100, 164; theory, 92; values and, 78; wind energy and, 88–96
Covenant on Economic, Social and Cultural Rights, 130
cows, infertility, illness and death in, 64–65
coyote (middle man), 56–58
crime, 69–70, 159
critical agrarian studies, 16–17
crop-insurance settlements, 39–40
Las Cruces, 32
Cruz, Melquiades, 81
Cruz Velázquez, Lucila Bettina, 128, 134–35, 139
CSR. *See* corporate social responsibility
CTF. *See* Clean Technology Fund
Cue Monteagudo, Gabino, 44
cultural change, 158–61
cultural genocide, 100–101, 153–66, 175
cultural heritage, 93
cultural identity, 103
cultural integrity, 161
cultural practices, 18, 24
cultural values, 166
cultures, 1–2, 11

death squads, 83–88
Demarest, Geoffrey, 81
democracy, 114, 144–45, 151
demonstrations, *xxvii*, xxviii
deregulation, xxiii
Diaz, Porfirio, 34
Díaz Ordaz, Gustavo, 37
Dirty War, xxv
disappearances, xxi–xxii, xxvii
discrimination, 73n5
dispossession, 138
division of labor, 150
dizziness, 71, 98, 170
doubly fed induction generator (DFIG), 99
drought, 61
drugs, xxi, 4–6, 69–70, 73n8, 89, 159, 164, 170

Drug War (Mexico), xxi
drying, 97–98, 157
duality, xxiii
Dugger, William, 10
dystopia rebranding, 172–78

ecocide, 3
ecological catastrophe, 3
ecological degradation, 3
ecological impact, 65, 97, 156–58
economy: asymmetry of, 25; green, 99, 139, 143, 164, 166, 175; growth of, 70; incentive structure of, 3; industrialism and, 176; liberalization of, xxiii; of Mexico, 70; political, 154; restructuring of, xxii, 53; structural adjustment laws of, xxiii
ecosystem, disruption to, 18
education, xxii, 66, 118, 162. *See also comunalidad* education
EIA. *See* environmental impact assessment
ejidos. *See* communal land
elections, 35, 38, 115–17, 120, 122
electricity: extraction of, 44; free, 52; in ground, 64; infrastructure for, 45; noise of, 100; prices, 157; profitable, 2–3; rates, 66–67; supply grid, 53
elites, 3
employees, 156
employment, 1, 11, 66, 70, 82–83, 92, 124–25, 160
Energy and Utility Act, 81
environment, 12, 91–92, 119
environmental ethics, 76
environmental impact, 50–53, 71, 97, 108–9, 119, 121, 138, 158
environmental impact assessment (EIA), 139–41, 143
environmentalism, market-based, 148
Eólica del Sur. *See* Mareña Renovables wind park
EPR. *See* Popular Revolutionary Army
equalitarizaton, 31
equal rights, 130
Equator Principles, 130
ethics, 169
ethnic cleansing, 163

ethnocide, 148
Eurus, 44
exploitation, 12, 33, 128, 138
Export-Import Bank, 52
extermination, 100, 146n1, 148, 153, 155, 163, 167. *See also* genocide
extractivism, 14, 20, 177

FAO. *See* Food and Agriculture Organization of the United Nations
farming, 62, 149, 157, 161–62, 163–64
FDI. *See* foreign direct investment
Federal Electricity Commission (Comisión Federal de Electricidad) (CFE), xxii, 43–44, 51
Federal Electric Utility Act, 51
Federal Preventive Police (PFP), 79
fertilizers, 161–62
finance, *173*, 174
financial realism, 130, 141
fish, 96–97, 108–10, 118, 158, 163
fishermen, 92, 108–9, 112, 158
flooding, 97–98, 157, 164
food, 14, 68, 70, 121, 160–62. *See also* agriculture
Food and Agriculture Organization of the United Nations (FAO), 18
foreign direct investment (FDI), xxiii, 1, 50, 139, 141, 164–65
foreigners, 160, 161, 167n7
foreignisation, 18
foreign workers, 167n6
forestry, 82, 101n10
fossil fuel industries, 18, 149, 157, 159, 174, 175, 176–77
fossil fuels, 26, 99
Foucault, Michel, 13, 20–21, 153
Free, Prior and Informed Consent (FPIC), 2, 21–22, 73n5, 88, 96, 101, 121, 131; advertising about, 141; APPJ and, 134; boycotting, 145; consultation, 50, 127–29, *129*, 133–45, 171; counterinsurgency and, 145; democracy and, 144; discursive techniques by, 25; EIA, 139–41, 143; free aspect of, 136–37; health complications and, 140; IIED and, 132; indigenous autonomy and,

129–30, 145; indigenous rights and, 138, 143; in Isthmus of Tehuantepec, 127; in Juchitán, 157–58; justice and, 132; land deals and, 132; legitimizing development and, 141–44; mainstream popularity of, 133; at Mareña Renovables wind park, 127–28; policy, 132; political boss and, 136–37; political parties and, 134; prior and informed aspect of, 137–41; process, 171; standards, 129–31; TC, 128–29, *129*, 133, 135, 136, 138–39, 143–44; technical violations of, 141; violence and, 128, 136, 144
fuel consumption, 77

Gamesa, 50
Gas Natural Fenosa (GNF), 1, 88, 92
Gato, Elder, 65
Gato Gris, 59
gender, 23, 27n12
General Electric Wind, 50
General Law of Climate Change (LGCC), 2–3
Geneva Conventions, 43
genocide, 3, 11; capitalism and, 154; colonial, 148, 152–56; complexity of, 153; consequences of, 156; counterinsurgency tactics as, 155; cultural, 100–101, 153–66, 175; Foucault and, 153; indigenous people and, 155, 175; industrial, 148, 166–67, 171–72; Lemkin on, 152–53; machine, 154; origin of word, 152; politics and, 167n2; slow, industrial, 25; violence and, 155; wind turbines and, 148
genocide-ecocide nexus, 163–65
gentrification, 24, 49, 67–71
Geographic Information Systems (GIS), 81
geography, 16
GIS. *See* Geographic Information Systems
globalization, 155, 178
Global North, 52
Global South, 15, 52
GNF. *See* Gas Natural Fenosa
God, 167n11
Gomez, Che, 24
Gomez, Jose F., 34

Gómez, Sergio, 18–19
governance techniques, 124
government. *See* Mexican government
grants, anthropology and, 14
Greece, 18
green: economy, 99, 139, 143, 164, 166, 175; extermination, 100; grabbing, 17, 20, 124, 163, 171; industrial development, 50–53; militarization, 76; uranium, 52; violence, 76–77; wind energy, 72
greenhouse gas emissions, 142, 176
Green Peace, 176
Green Rush, 50
grounding, of wind turbines, 64
Grupo Mexico, 53
Guela Venge, 86
Gui'Xhi' Ro. *See* Álvaro Obregón
guns, 8, 43, 115–16, 125n5, 137
Güven, Ferit, 151
de Gyves, Leopoldo, 38–39

habitat, wind turbines and, 63–64, 97, 160, 163
harassment, 110
health impacts, 11, 24, 71, 96, 98, 140, 142–43, 170
hierarchy, 150
The Higher Learning in America (Veblen), 10
high voltage transmission lines, 104
Hoffman, D., 15
Holocaust, 153, 154
'home-field' dichotomy, 10
household governance (*oikonomia*), 151
House of Culture, 84, 94, *129*
Huave, 76
human rights, xxi, 4, 77, 90, 95, 103, 115
hunger, 121
hunting, 159–60
hydraulic fracturing, 25, 149, 172
hydroelectric projects, 19, 42

Iberdrola, 44, 50, 66
ICJ. *See* International Court of Justice
identity, 16, 86, 132
IFC. *See* International Financial Corporation

IIED. *See* International Institute for Environment and Development
Ikoots, 29, 43
illiteracy, 56
ILO. *See* International Labour Organization
imperialism, colonialism and, 149
imports, food, 70
impunity, xxi, xxiii
IMSO. *See* Integrated Monetary Shaping Operations
inclusionary control, 145
income, 58, 67, 70–71, 143, 144, 157, 159, 164–65
Independent Power Production (IPP), 43
India, 18
indigenous: autonomy, FPIC and, 129–30, 145; colonial history against, 146n1; communities, xxi, 2, 15, 18, 29–31, 114, 147–48; cultures, 100; customary laws, 113; genocide and, 155, 175; governance, 7; identity of, 132; as label, 166; languages, 29; resistance movement, 7; re-vitalization, 11–12; rights, 130, 138, 143, 171; self-determination, 131, 144
Indigenous and Tribal peoples Convention, 130, 139, 171
industrial-capitalist system, 3
industrial degradation, 99, 164
industrial development, 3, 76, 166, 175, 177
industrial genocide, 148, 166–67, 171–72
industrialism, 26, 172, 176, 178
Industrial Presence and Commercial Police (PABIC), 88–89
industrial revolution, 148
industrial waste, 175
industrial wind turbines (IWTs), 27n8
INE. *See* National Electoral Institute
inequality, 49, 123
information, 90
infrasounds, 140
infrastructure, 45, 65, 99, 159, 178
Institutional Revolutionary Party (PRI), xxiv, 5, 35–38, 40, 55, 106, 107, 118–19, 134
Instituto de Investigaciones eléctricas, 44

insurgency, 78, 79, 80, 144, 145. *See also* counterinsurgency
insurrection, caciquismo and, 33–37
Integrated Monetary Shaping Operations (IMSO), 91, 92–95
intelligence networks, 77
intent to destroy, 156
Inter-American Commission on Human Rights, 115
inter-communal conflicts, 5
Inter-governmental Panel on Climate Change (IPCC), 23–24
International Court of Justice (ICJ), 130
International Financial Corporation (IFC), 1, 130–31
International Institute for Environment and Development (IIED), 132
International Labour Organization (ILO), 21, 127, 130, 171
international regulatory boards, 133
International Renewable Energy Agency (IRENA), 50
intimidation, 56, 60, 136, 144
IPCC. *See* Inter-governmental Panel on Climate Change
IPP. *See* Independent Power Production
IRENA. *See* International Renewable Energy Agency
isolation tactics, 96
Isthmus Coalition of Workers, Peasants, and Students (COCEI), 24, 35–40, 47n7, 80, 106–7, 118–19, 125n6, 134, 141, 146n3; activists for, 54–55; position of, 42; support in Álvaro Obregón, 107–8
Isthmus of Tehuantepec (*Istmo*), xxiv–xxix, 76, 105; communal land of, 106; cultural change in Northern, 158–61; culture of, 119; FPIC in, 127; genealogy of, 45; history of, 24, 29–31; map, *30*; North and South, 156–58; social property in, 80; Southern, 161–62; wind energy generation and, 107; wind parks in, by year, *46*; wind turbines in, 107
Istmeño: Viento de Rebeldia (film), xxvi
IWTs. *See* industrial wind turbines

jobs. *See* employment
Jormungand, Hex, 147
Juárez, Benito, 33
Juárez-Hernández, Sergio, 44
Juchitán, 23, *30*, 33, 39, 83, 84, 118, 157–58
juridico-political system, 31

Kay, Cristóbal, 18–19
Kenya, 18
kidnapping, xxii, 5
Kitson, General, 117
knowledge, 9–15

labor, xxi, 156
Labor Party (PT), 41
Laguna Superior, *30*, 77, 103, 104, 127
land: access, politics of, 53–58; acquisition, 24, 49, 131, 145; change, 14, 65–71; concentration, 18; contracts, negotiating, 58; control, techniques for, 56–57; deals, 156; deals, FPIC and, 132; displacement, 17; grabbing, 4, 17, 18; leasing, 19, 68, 84–85; manipulation, 157; owners, 3, 84; titling, 47n1. *See also* communal land
La Venta wind park, *30*, 52, 157
La Ventosa, 1, 21–23, *30*, 45, 49, 52, 55–56, 67, 69–70, 72, 157, 159–60
La Ventosa Vive group, 59
laws, Indigenous customary, 113
legal status, of communal land, 83
legislation, neoliberal, 124
legitimacy, of wind energy development, 141–44, 171
Lemkin, Raphael, 152–53
León, Gabriel, 44
LGCC. *See* General Law of Climate Change
liberalism, 154
lifestyle, 159–60
lightning, 63
livestock, 70
local gunmen, 88
local medicine, 97
López Castellanos, Miguel, 6, 107
López Nelio, Daniel, 38
Low Intensity Operations (Kitson), 117
LyFC. *See* Central Light and Power

manipulation, 58, 157, 166
Mapache, 163
mapping, 81–82
Mareña Renovables wind park (Eólica del Sur), 2, 5–6, 25, 45, 104–5, 107, 164–65, 170–71; Barra and, 110–16; barricade, 110–12, 127; climate change and, 123; communal land and, 110; contract, 122; fishermen on, 108–9; FPIC and, 127–28; neoliberal ideology in, 165; new location of, 116, 121; political parties and, 113, 117; roads and, 121–22; security guards of, 109, 111
marginalization, 24
Marker, Michael, 11
marketing, *173*, 174
market mitigation, 72
marriage, 57–58
Martínez-Alier, Joan, 17
La Mata wind park, 52
Matus, Oscar, 38
meat prices, 68
media, xxviii, 77
medicinal plants, 85, 163
medicine, 97, 161, 163
Melendez, Jose Gregorio, 33
mental health, 140
mercenaries, 88–89
Mérida Initiative, xxii
Mexican Communist Party, 38
Mexican Electrical Workers Union (SME), 51
Mexican government, 81, 114, 117, 139, 148, 163
Mexican National Forestry Commission (CONAFOR), 82
Mexican Revolution, 34, 105, 134
Mexico, *30*, 33–37, 50, 70, 79–83
México Indígena, 81, 92
M-Form corporate model, *173*, 174–75
micro-politics, 16, 21
middle man (*coyote*), 56–58
militancy, 9–16, 40
military, xxi, *xxviii*, 12, 14, 76–77, 81, 100, 149
mining, 52

Mixes, 29
modernization, 35, 40, 106, 119, 147–48, 154, 160, 164
monetary shaping operations, 93
money, 91, 140
Monjardin, Adriana López, 45
Monsanto, 47n5
Monte Albán, 31
Moses, A. Dirk, 150
Mother Nature, 167n11
mothers, 70
Movimiento de Unificación de las luchas Triquis (MULT), 101n12
municipality takeover, 110–16
mural, *69*
murder, 153
Musalem, Manuel (Tarú), 37
My Secret Life (Anonymous Author), 13

NAFTA. *See* North American Free Trade Agreement
narcotics. *See* drugs
National Electoral Institute (INE), xxiv, 115
national parks, militarization of, 17
National Renewable Energy laboratory, 41
National Solidarity Program, 41
Native North American resistance, 100
natural resources, 76, 81
nature, 91–92
Nazi regime, 11, 153
neocolonialism, 3, 176
neoliberalism, 41, 53, 100, 113–14, 124, 148, 154–55, 160–61, 164–65, 175
New Agreement for Indigenous Peoples. *See Nuevo Acuerdo para los Pueblos Indíenas*
NGOs, 90, 177
NIMBY. *See* 'not in my back yard'
noise, 98, 109, 140, 158, 170
non-profit organizations, 19
non-violent social movements, 79
normalization, 142
North American Free Trade Agreement (NAFTA), xxii, 40, 51, 79
'not in my back yard' (NIMBY), 17
nuclear energy, 25, 98, 149, 172

Nuevo Acuerdo para los Pueblos Indíenas (New Agreement for Indigenous Peoples), 114

Oaxaca, xxii, *30*, 80, 83, 111
Oaxaca Insurrection, 43, 83
oikonomia. *See* household governance
oil, 25, 39, 62, 98, 109, 140, 149, 172
Olivas, Ramon, 141–42
Olympic Tlatelolco massacre, 37
OPORTUNIDADES, 81
oppression, 15–16, 45
out-migration, 70–71, 157, 160, 162, 164, 170

PABIC. *See* Industrial Presence and Commercial Police
Pacto de Concertación Social. *See concertación* accords
Pan American highway, 36
paramilitaries, 4
paranoia, 90
Parques Ecológicos de México, 44
participant observation, 16
Party of the Democratic Revolution (PRD), 40, 47n7
paternalism, 150
patriarchy, 14, 120
payment for ecosystem services (PES), 82, 121
PEMEX oil company, xxii, 37, 51
Peña Nieto, Enrique, xxiii, 35, 51
Peñoles, 53
People's Assembly of San Dionisio del Mar (ADPSM), xxiv, 5–6
people's government (Ayuntamiento Popular), 39
Perceptions of the Wind Energy Park in Juchitán, Oaxaca, 91
PES. *See* payment for ecosystem services
Petroleum Act and Federal Electric Utility Act, xxii
PFP. *See* Federal Preventive Police
Philips, Martin, 67
Pike, Fredrick, 31
pilgrimages, 86, 95
Plan Puebla Panama, 53

plants, 97, 163
Plato, 151
Playa Vicente, 21
police, xxi, xxviii, 4; anthropologists working with military and, 14; brutality, 14, 38, 43; controls, 110; interventions, 111; occupations, xxiv–xxv; patrols, 111; violence, 111
policía comunitaria. See Communitarian Police
Polis, 151
political affinities, 16, *69*
political boss (*cacique*), 54–55, 60, 71, 136–37
political control, 151
political culture, 100
political ecology, 16, 20
political economy, 154
political identity, 153
political parties, 7, 113, 117, 134, 163
political representatives, 83
political repression, 22
political unity, 151
political violence, 155
politicians, 3, 6
politics, genocide and, 167n2
pollution, 85
Popular Assembly of the Juchiteco People (APPJ), 75–76, 86, 87, 90, 92, 97, 134
Popular Assembly of the Peoples of Oaxaca (*Asamblea Popular de los Pueblos de Oaxaca*) (APPO), 43
Popular Revolutionary Army (EPR), 79
population control, 81, 117
positionality, knowledge, militancy and, 9–15
poverty, 49, 67, 79, 96, 121, 124, 128, 140, 162
power asymmetries, 131
practices and customs (*usos y costumbres*), 7
PRD. *See* Party of the Democratic Revolution
PRI. *See* Institutional Revolutionary Party
prison system, xxi
privatization, xxii, 18, 51, 85, 106

PROCEDE. *See* Program for the Certification of Ejido Rights and Titling of House Plots
processing facilities, 159
production, *173*, 174
profitability, 50
Program for the Certification of Ejido Rights and Titling of House Plots (PROCEDE), 34, 80–81
PROGRESA, 80–81
PRONASOL, 80–81
property, 80–81, 82, 106
prosperity, 1, 59
protests, *xxviii*, 75–76, 78, 87
proxy techniques, 124
psychological operations (PSYOPS), 78
PT. *See* Labor Party
public good, 14–15
public notaries,l communal land and, 85

quality of life, 65–66, 161

racism, 12, 59, 73n5, 164
radio, 5
radioactive waste, 99
rain, 85
Ramírez, Gustavo, 81
Rasgado, Federico, 106, 109
recycling, 175
Regalado Jiménez, Hector, 89
regulations, xxiii, 64
Reina, Letica, 34
renewable energy, 2, 23, 104, 148, 172, 175–76, 177, 178n1
rent, 68
repression, xxii, xxviii
Republic (Plato), 151
resistance, xxiv, 88, 156, 166, 171; in Álvaro Obregón, 9; against Bíi Hioxo wind park, protests and, 75–76, 78, 87; colonial incursion, adaptation and, 29–33; types of, 138; against wind energy development, 3, 43; to wind parks, 1–2, 58; from Zapotecs, 43
right of conquest, 150
Rios, Manuel, 55
rituals, 103

roads, 61, 66, *87*, 87–88, 97–98, 121–22, 135, 149
Rome, 31, 149
Rubin, Jeffery W., 36, 38–40, 106
Ruiz Ortiz, Ernesto Ulises, xxii, 43
rumours, 90
rural gentrification, 67–71, 157

sacred sites, communal land and, 86
safety, 91
Salinas de Gortari, Carlos, 40–41, 47n7, 106–7
San Andrés Accords negotiations, 114
Sánchez, Hector, 38, 40–41
Sánchez López, Héctor, 106–7
San Dionisio, 104, 157
San Mateo, 104, 157
Santiago, Lorenza, 38
SAPs. *See* structural adjustment programs
Scheper-Hughes, N., 15
School of the Americas (SOA), 79
schools, 4, 5, 66, 118, 161–62
Scotland, 18
Secretary of Indigenous Affairs, 92
securitization, of universities, 11
security cameras, *87*
security guards, 86, 109
security provision, 77
self-censorship, 61
self-destruction, 155
self-determination, 34
self-management, 155
'self-supply' (*autoabastecimiento*), 44, 50
settler colonialism, colonialism and, 150, 160
sexual abuse, 167n6
sexual deviance, 13
sexuality, 13
ship building, 32
Sierra Atravesada mountain range, 49
Sierra Club, 176
silver, 32
slavery, 32, 119, 149, 162, 176
SME. *See* Mexican Electrical Workers Union
Smith, Darren, 67
SOA. *See* School of the Americas
social benefits, 65–66

social change, 13
social conflict, 133
social control, coercion and, 155
social death, 153
social development, 1, 49, 59, 77, 92–94, 93, 156–57
social division, 131
social impacts, 71
social movements, 79, 91, 176
social networks, 92
social norms, 26n3
social property, 80, 106
social warfare, 20–21
social welfare, 80–81
solar energy, 20
solidarity caravan, *xxvii*, xxix, 4
Sosa, Alvaro, 107
Spain, 18, 33
Special Climate Change Program, 2, 3
specialization, 150, 167n6
speculation, 92
sports, 66
State Agency of Investigations (AEI), xxiv
structural adjustment programs (SAPs), 40
students, disappearances of, xxi
subsidies, 66–67, 142, 161, 167n8
sugar cane, 55
suicide, 175
sun, 2–3
sustainability, 51, 76, 82, 104, 156, 169
Sweden, 18

Tarú. *See* Musalem, Manuel
Technical Committee (TC), 128–29, *129*, 133, 135, 136, 138–39, 143–44
Tehuantepec Rebellion, 1660, 32
temporary employment, 66
Tepeyac Center for Human Rights (CDHT), 42
Third Reich, 11, 153
tidal wave energy, 20
timber concessions, 19
tourism, xxv
town square mural, *123*
tradition, 114, 117–18
Transcendence (film), 176
Trans-Isthmus highway, 36

Trump, Donald, 23–24
turbines. *See* wind turbines
Tutino, John, 31–32

UCIZONI. *See* Association of Indigenous Communities in the Northern Zone of the Isthmus
U-Form corporate model, *173*, 174
UN Convention. *See* United Nations Convention on the Prevention and Punishment of the Crime of Genocide
UNDRIP. *See* United Nations Declaration on the Rights of Indigenous Peoples
unemployment, 19, 170
UNFCCC. *See* United Nations Framework Convention on Climate Change
unions, 59
United Nations Convention on the Prevention and Punishment of the Crime of Genocide (The UN Convention), 152–54
United Nations Declaration on the Rights of Indigenous Peoples (UNDRIP), 130
United Nations Framework Convention on Climate Change, 2012 (UNFCCC), 2
United States (US), 18
Universal Declaration on Cultural Diversity, 130
universities, 11, 12
US Agency for International Development (USAID), 47n8, 107
usos y costumbres (practices and customs), 7

values, counterinsurgency and, 78
Vásquez, Genaro V., 35
Veblen, Thorstein, 10
Velas festivals, 94–95
vertigo, 98
Vicente, Saúl, 60
Vidal, Gore, 17
violence, xxi, 99; in Álvaro Obregón, 122; bureaucratic, 133; Communitarians and, 115–16, 120; drugs and, 73n8; epistemic, 12; FPIC and, 128, 136, 144; genocide and, 155; green, 76–77; intimidation and, 60; paramilitary, xxviii; police, 111; political, 14, 155; state, 128; structural, of universities, 12; types of, 9
voting, 115

wage slavery, 162
Wales, 18
Walmart, 51, 53
war hypothesis, 21
Warman, A., 37
War on Drugs, 79, 80
water, 64, 77
welfare, 118–19
Wilkinson, John, 18–19
The Will To Knowledge (Foucault), 13
wind, 2–3, 4, 68, 85
wind energy, 20; agriculture and, 70, 142; arrival of, 53–58; climate change and, 123; Communitarians and, 119; companies, xxiv, 56, 65, 83, 107–9, 139, 148, 156; counterinsurgency and, 88–96; export of, 52; genocide-ecocide nexus and, 163–65; impact of, 4; Isthmus of Tehuantepec and, 107; manufacturing, 177; projects, xxiv, xxvi, 18, 24, 42, 50
wind energy development, xxiii, 45, 104, 124, 162–63; capitalism and, 174–75; civil conflict and, 113, 138; climate change and, 149; as dystopia rebranding, 172–78; fossil fuel industries and, 18; legitimacy of, 141–44, 171; modernization and, 164; natural medicine and, 163; political ecology of, 20; political repression for, 22; rejection of, 156; resistance against, 3, 43; as skyrocketing, 43; social impact of, 21; structural problems with, 3
Wind Energy Resource Atlas of Oaxaca, 43
wind parks, 93; construction of, 21–22, 50–53, 63; defining, 44; emergence of, 44; employment opportunities in, 57–58; environmental impact of, 50–53; growth of, 156–57; in Isthmus of Tehuantepec by year, *46*; phases of, 44–45; political parties and, 7; resistance to, 1–2, 58. *See also specific parks*
Winds of Change (exhibition), 93–95, *95*

wind turbines, xxiv, 4, 20, 47n9; animal mortality and, 62–64, 96–97, 163; in Barra, 128; of Bíi Hioxo wind park, 77, 83–88; colonization, 22; construction of, 99, 125n8, 158, 159; contracts, 84; cultural genocide and, 156–67; ecological impact of, 65, 97; environmental impact of, 158; farming and, 157, 161–62, 163–64; fire, 63; fish population and, 96–97, 108–9, 118, 158, 163; foundations, 97, 109, 157, 158; genocide and, 148; grounding of, 64; habitat and, 63–64, 97, 160, 163; health complications and, 71; in Isthmus of Tehuantepec, 107; lights on, 96, 158; locations of, 17; modernity and, 162; oil leaking from, 62, 98, 108; public stance against, 96; slavery and, 162; sustainability and, 169; in La Ventosa, 49, 72

wind turbine syndrome, 71

women, Communitarian Police and, 120
World Bank, 41, 52
World War I, 173
World War II, 174

youth, 92

Zapatista movement, 82, 114
Zapatista uprising, 41
Zapotecs, 29–32, 42, 76; cultural life of, 86, 93–95, *95*, 100; cultural sites, 85–86; culture, 106; elites, 37; etymology of words in, 101n14; identity of, 39; religion of, 85; resistance from, 43; rituals, 103; sacred sites, 86; spelling of words, 101n13; tombs, 86; *Velas*, 94–95
Zeitlin, Judith, 31–32
Zenteno, Eduardo, 135
Zografos, Christos, 17
Zoques, 29

www.ingramcontent.com/pod-product-compliance
Ingram Content Group UK Ltd.
Pitfield, Milton Keynes, MK11 3LW, UK
UKHW041832220426
470268UK00001B/26